Human Minds and Cultures

Sanjit Chakraborty
Editor

Human Minds and Cultures

Springer

Editor
Sanjit Chakraborty
School of Social Sciences and Humanities
Vellore Institute of Technology-AP University
Amaravati, Andhra Pradesh, India

ISBN 978-3-031-56447-5   ISBN 978-3-031-56448-2 (eBook)
https://doi.org/10.1007/978-3-031-56448-2

This Springer imprint is published by the registered company Springer Nature Switzerland AG
The registered company address is: Gewerbestrasse 11, 6330 Cham, Switzerland

Paper in this product is recyclable.

This book is dedicated to my parents—
Mrs. Pushpa Chakraborty and
Mr. Sunil Chakraborty, inspirations of my academic quests!

Preface

Aiming for multicultural spectrum and philosophical explorations that could help us reflect on the pursuit of envisaging human minds and cultures as a decisive panorama of philosophy and social sciences, the standalone thematic volume titled *Human Minds and Cultures* presents a harmonious analysis of the similarities and differences between two concepts: human minds and cultures. The book rests on a theoretical account of the essential relationship between the dualities of minds and matters found in semantic-ontological conjectures. Cultural scientism, an evolutionary scaffold on the functionalist shift, is limited in its extent by the ethnographic understanding of cultures, which allows one to envision cultural phenomena from the resolution of scientific conjectures. Thus, cultural relativism acknowledges that cultures include descriptive elements that individuals or groups of people adhere to, such as customs, beliefs, moral standards, and other minds. Besides, the idea of a "semiotic conception of culture," which centers on the idea of ethical theorizing, employs modernism and values in its vision of social semiotics. *Human Minds and Cultures* remarkably aligns with the cultural structure of moral minds and addresses the ever-present issues of gender, law, ethics, and sociopolitical events. For anyone with a keen interest in minds and cultures, this is an essential reading.

So here is the book, and it's been amazing putting it together. Too much bogged down in diversified academic and nonacademic commitments sparks thoughts that it would have been a more engaging volume if the writings of a couple of more distinguished thinkers had been included at the last moment. The authors who have contributed to the volume are stellar in their fields. Professor Harry Gensler passed away before the publication of this volume, and we greatly missed him. I'm quite indebted for what he added to the volume.

Other intellectual obligations go to my good friends and mentors, the late Hilary Putnam, Noam Chomsky, Amartya Sen, Martha Nussbaum, Frank Jackson, Jan Narveson, Deborah Poff, Duncan Pritchard, Robin Le Poidevin, Jeffery Long, and many others, for their philosophical déjeuners.

Mr. Christopher Wilby, my editor at Springer Nature, is much more than just a professional associate, and I amenably thank him and other Springer team members for everything they have done for the book. I am grateful for the supportive and encouraging academic environment provided by my colleagues at VIT-AP University and my former institution, IISER Kolkata.

Amaravati, Andhra Pradesh, India Sanjit Chakraborty

Contents

Human Minds and Cultures: An Introduction ... 1
Sanjit Chakraborty

Thoughts on Morality and Culture ... 19
Jan Narveson

Values and Cultures ... 29
Harry J. Gensler

Clashes of Culture ... 73
Michael Smith

Hatred: Why Do Such Nice People Do Such Awful Things? ... 89
Michael Ruse

Neoliberalism, Globalization and Discrimination in the Twenty-First Century: Issues, Challenges and Limitations of Habermas' Reason-Based Theory of Liberal Democracy in Current Times ... 109
Deborah C. Poff

Legal Emotions: Why Affect Matters for Law ... 123
Antara Haldar

The Epistemic Argument for Hedonism ... 137
Neil Sinhababu

The Interpreting Mind and Meaning-Formation: A Relational Critique of Lyotard and Ricœur ... 159
Gerald Cipriani

Recurring Dilemmas on Other Minds ... 175
Sanjit Chakraborty

References ... 189

Contributors

Sanjit Chakraborty School of Social Sciences and Humanities, Vellore Institute of Technology-AP University, Amaravati, Andhra Pradesh, India

Gerald Cipriani Graduate Institute of Philosophy, National Central University, Taoyuan City, Taiwan

Harry J. Gensler Loyola University, Chicago, IL, USA

Antara Haldar University of Cambridge, Cambridge, UK

Harvard University, Cambridge, MA, USA

Stanford University, Stanford, CA, USA

Jan Narveson University of Waterloo, Waterloo, ON, Canada

Deborah C. Poff Carleton University, Ottawa, ON, Canada

Michael Ruse Florida State University, Tallahassee, FL, USA

Neil Sinhababu National University of Singapore, Kallipolis, Singapore

Michael Smith Princeton University, Princeton, NJ, USA

Human Minds and Cultures: An Introduction

Sanjit Chakraborty

> *It is true that you can't make an omelette without breaking eggs; but you can break a great many eggs without making an omelette.*
>
> –Hannah Arendt

This thematic volume entitled ***Human Minds and Cultures*** deciphers different aspects of human minds and cultures that differ in their methodological patterns. It exhibits humanity as a universal, underlying cultural multiplicity coping with diverse prospects of normative morality, semiotics, and socio-linguistic human affairs. Two major concerns that the thematic volume anticipates here are as follows:

First, an assessment of literature is traced back to human affairs and cultures. Literature transmits human life into artistic conventions where cultures and minds echo their intermingled alignments with the subjective and inter-subjective panorama of morality, understanding, and knowledge. The principle of "incommensurability of cultures" revives the idea of the universal comprehension of humanity, which is infused with the dimensions of moral values in Charles Taylor's words and "accord equal respect to actually evolved cultures" (Taylor, 1994: 42). This doubtlessly seems a contentious issue. As I have written elsewhere, "Besides, the context of the abiding expansion of universal-realization of humanity needs to be linked with the dimensions of moral values, care, and philosophizing life — an art of life with numerous ecosystems" (Chakraborty, 2021b: 774).

Second, a sociocultural value is inferred from the author's inter-subjective intention and the conscience articulated by their intellect, liberal ideas, and remarkable praxis in society. This technique is not a social artefact but a dynamic insight in which moral philosophy preserves someone and endows those thinkers with their moral parameters, freedom, and the notion of choices. The concept of humanity is

S. Chakraborty (✉)
School of Social Sciences and Humanities, Vellore Institute of Technology-AP University, Amaravati, Andhra Pradesh, India

S. Chakraborty (ed.), *Human Minds and Cultures*,
https://doi.org/10.1007/978-3-031-56448-2_1

surplus in creativity, and human minds intermingle with others and the world through freedom, love, and societal norms. In Keats' sense, "soul-making" is nothing but the celestial relation between the poet and poetry, or the extended mind and nature. Minds outpace the non-naturalistic transcendental sphere by paving the way first to subjectivity and then to morality. The epistemic representation of the mind precedes an ontological explanation of the mental contents.

Let me introduce an overview of the epistemic structure of thought in an effort to revitalize human minds and mental contents. To know what it means to be "economically challenged" and how a poor person survives in society, you have an opportunity to live like a poor person and experience their everyday wretched situations. But to know the conception of non-thought or the mode of non-thinking, a human mind cannot enter within the sphere of a thoughtless or mindless state. The structure of thought is a spontaneous continuation of our mental representation and a task of making the mind think. The point is badly framed since we have to be aware of the part of speaking (not only in a verbal sense but in the sense of internal speaking). We think in the mode of language, and the socio-linguistic structure helps us to construct our thoughts. Wittgenstein argues, "Remember that our language might possess a variety of different words: one for 'thinking out loud'; one for thinking as one talks to oneself in the imagination; one for a pause during which something or other floats before the mind, after which, however, we are able to give a confident answer" (Wittgenstein & Anthony, 1994, 124).

Considering the phenomenon of thinking, I shall begin by discussing the view that thinking is a conscious activity of minds that controls a subject's physical actions. A dualistic plea emerges when we are intuitively drawn to a dualism of mind and body by probing the concept of thought. Assigning concepts to the mode of thinking opens a psychological question on the study of epistemic semblance in ontology. As we know, Quine's inclinations to a naturalistic conception of knowledge anticipated the "knowledge of knowledge" method. Epistemological queries have two diverse facets. The "conceptual side" of conventional epistemology sketches the different ontological concepts of phenomenal knowledge. Besides, the "doctrinal side" contends with the justification that validates those concepts, which are exposed in the "conceptual side" of epistemology (Quine, 1969, 69–70). Quine articulates that the "conceptual side" looks analogous to the theory of meaning, while the "doctrinal side" also copes with the theory of truth and reference. Kim offers a critical discussion and argues, "In urging naturalized epistemology on us, Quine is not suggesting that we give up the Cartesian foundationalist solution and explore others within the same framework-perhaps to adopt some sort of 'coherentist' strategy, or to require of our basic beliefs only some degree of 'initial credibility' rather than Cartesian certainty … Quine's proposal is more radical than that. He is asking us to set aside the entire framework of justification-centered epistemology. This is what is new in Quine's proposals. Quine is asking us to put in its place a purely descriptive causal-nomological science of human cognition" (Kim, 2008, 304–305).

To get into the causal nomological science of the human mind and cognition is nevertheless an attempt to decipher mental contents from the scheme of externalism. One could think that the mental content of one's thought is an act from objects to conceptual schemata that are saturated by truth values. An incongruity between meanings and mental contents may be feasible as sentences with varied meanings can resolve the same functions. In addition, to begging the crucial question, it sounds as if there is an impending condition in which all thoughts do not have a causal efficacy on the world. There are some thoughts that hold liaison to the world; these are states of the mind with underlying causal and external backgrounds that engage in many factors including the world, evolutionary forces, experience, etc. Besides, some mental contents are not conjoined to the matter of the facts or causal efficacy, and these pivot idiosyncratic efforts on agents' behaviours. For instance, in some cases where mental contents are preceded by physical objects like in the case of deaf and dumb children, here mental construction of thoughts can perpetuate without any physical inputs.

An outline of mental content methodologically strengthens a paraphrase of normative and representational accounts to bring an assessment of the philosophical understanding of the mental representation of thoughts. The various criteria of content deploy the systems of mental representation as follows:

(a) *Content-laden intentionality*:

Content that in actuality refers to mental content puts emphasis on a criterion of being possession of intentionality just is having aboutness. One should not get the wrong impression about the conception of intentionality with the emerging notion of intension(s), a propositional content of an agent's cognitive attitudes. I wrote elsewhere, "I think it would be interesting if we think of mind as inner that has some outer properties like intension, action, and effects. All these symptoms or properties of the mind have some tendency to merge with the outer world. The naturalistic outer world that rebuts the mind as something inner cannot be the right opinion. The agents' thoughts, imaginations, and emotions are beyond communicative language, so we can put all these in the realm of inner entities about which the subject has a sort of privileged access. For me, the innerness of mind can be regarded as privileged in nature. It can only be expressed in the public domain when a subject structures their thoughts/sensations in a sentence to give it a linguistic form that could be shareable with others in our linguistic community" (Chakraborty, 2020, 211–212).

The philosophical jargon arises when we put the term "intention" that possesses the state of intentionality in a wide framework to incur an ability to be about something objective. We can put it in another phrase by saying that an intentional property of the mental state is a type of semantic property that emphasizes a regimented relation between a word and its referent. One could render a little more mental representational account to say that the fundamental structure of internal states of mental representation elevates a rudimentary sentential structure and both are analogous in a plausible sense. Well, I don't think that a non-semantic relation between mental representations and sentences of the communicative language may secure the causal relational formula of synonymy of contents. A difference in the causal relation to the mental content of the sentence entails a subsequent difference in the speaker's mental representation. We know about the famous objection of Jerry

Fodor and especially Stephen Stich who argue that it would be simply a mistake to spread out the view that a mentalese sentence shares a synonymous causal relation imbued in mental presentation with a public-language sentence. An inter-subjective sameness of understanding the conceptual role of mental contents could not be realistic across time and the contexts of the same speaker.

Let me articulate the thought using "that-clauses" since some philosophers (Stich, Fodor, and Wittgenstein) believe in mental representation as the referent of "that-clause". Thus, we can put the mental presentation in the form of a *propositional attitude* as in the following utterance:

S believes that "Hesperus is Phosphorus". As we know here the truth condition of the proposition is determined by its referential parts. If we infuse the truth condition of the components of the proposition by ramifying the conceptual role of mental content, we have to envisage a further co-related mental representation to ensure a delineation of the semantic value of the sentence with its tautological form "Hesperus is Hesperus". A fine informative approach of the primary belief (Hesperus is Phosphorus) and subsequent non-informative secondary belief (Hesperus is Hesperus) of a speaker does intuitively substitute a co-referentiality of the mental contents of the speaker to confine the truth value, but one might hold different cognitive values (mental presentations) in this mentioned case where the designator looks same as the planet Venus. So, we can indeed believe that the semantic power intends to some extra-linguistic entities, viz. meanings where the referent of a particular term signifies an object. The role of truth value entails an epistemic cum logical understanding of the *extensional context* in the Fregean sense where two substitutions of co-refereeing expressions barely alter their substituted truth values. For instance, if a proposition "A is M" is true and A = B, the proposition "B is M" is true too. Most of the case names designate persons that are entities of a special kind (natural kind terms); except for some restricted cases like natural kind terms, the rest of the other cases' references are not determined by the causal referential directness to the object but rather rely on the conceptual role semantic.

(b) *Normativity of content:*

The normativity of content highlights the propositional content and its cognitive values cum truth values where the subject can imply a term correctly/incorrectly or can consist of their beliefs/mental presentations in the context of truth values (true/false). Here a semantic power emulates a particular belief that is true or false. The claim of content-bearing phenomena along with the stance of normativity is illustrated in the writings of Wittgenstein, but it sounds contentious to claim that Wittgenstein induced the explanation of inductivism (or prediction on the future) in his thoughts. Wittgenstein asks, "A wish seems already to know what will or would satisfy it; a proposition, a thought, what makes it true – even when that thing is not there at all! Whence this *determining* of what is not yet there? This despotic demand?" (Wittgenstein, 1953, 437).

The model of normative content-laden structure deploys a kind of propositional attitude psychology or folk psychology, a common sense explanation and systematic understanding of human behaviours and mental presentations in retrospect of ontological explanation of a person and their tacit psychological attitudes. The prospect of the normativity of content relies on the norms of the subject's beliefs. Can an agent be logically and rationally infallible about their mental contents? A considerable dispute effects between qualitative characteristics of mental contents and the infallibility of the agents. Here the puzzle is whether the linguistic meanings of mental content can be known without knowing the semantic value of the particular content, like in the case of "I" or the content of "self-ascription". The subjective phrase "content of self-ascription", in my opinion, always refers to the individual using the term "I". Let's suppose that you receive a phone call and an unknown person is speaking over the phone, "I would like to nominate you as a distinguished professor of a particular university X". In this situation, you, the hearer who picked up the phone, might say, "Who are you?" Here, the hearer can perceive the utterance as they knows the meaning but they does not know the referent, except only "the utterer" of this utterance "I am hearing". It looks possible that a person may be familiar with the linguistic rules and apprehend the communication without knowing who uttered these words. So, in this case, the particular person is unaware of the semantic value (whom the word refers to) of the uttered term "I" of the anonymous speaker or phone caller. It is well known that the goal of analytical philosophy is to analyse and categorize ordinary concepts – also known as metaphysical concepts – as well as the mental contents that support philosophical riddles. The semantic value of a sentence is its *referent* or one may call it denotation which does not hold a relation to the truth value of the referred sentence. In the case of self-ascription or "I" here, we cannot get any semantic value of the term. I think that the same would be followed for the indexical terms like "this mental suffering will kill me very soon", etc. So, it may well be possible that we can know the linguistic meaning of a term without knowing its semantic value. I'll discuss this issue in more detail in the later section "Revisiting Thought", which includes an extended argument to synchronize content and thought in the perspectives of externalism and internalism.

(c) *Content structured:*

The systematic structure of content has two dimensions – logical dimension and ability. As we know Fregean analytic philosophy articulates the logical and rational structures of content to show that the patterns of mental states are in general logical and rational since the justification or truth value of the mental states is logically interconnected. Suppose you hold a belief (morning star is morning star) that could rationally justify holding another belief (morning star is evening star). However, the ability approach to understanding the mental states in the second level presupposes an interrelated constituent ability. In the model of understanding the mental representation, one has to understand the ability to mull over the propositional attitude first, and this structure of content will gradually align with the interrelated thoughts or propositional attitudes. One may put this in holistic assignments that nudge the structured content underlined by many others' systematic thoughts. Various logical terms including the well-known example "for the sake" by Noam Chomsky are linked to speakers' conceptual roles. But these conceptual roles and the apparatus have also relied on the external world in the case of understanding any particular term. As mental content is world-directed content, internal meaning can be fixed externally, especially in this case, where the environment plays a basic role.

Synchronizing Mind and Content

In this segment, my goal is to align the mental content by examining philosophical explanations and their heightened focus on reductionists, anti-reductionists, representationalism, and beyond. In examining the mental content in relation to the explanatory parameter, we differentiate between reductionist and anti-reductionist models. Given the explanatory priority to mental content or linguistic meaning, an attempt based on this strategy entails two different hypotheses:

(a) We can ascribe explanatory priority to the mental content and elucidate the way linguistic meaning follows from this.
(b) We can ascribe explanatory priority to the linguistic meaning and elucidate the way mental content follows from this.

While the reductionist account of content in this sense is genuinely non-intentional, it is closer to representationalism. It is founded on a content-directed hypothesis, where the necessary and sufficient conditions for any content would be concepts that are naturalistic and non-intentional. Tim Thornton clearly explains, "Fodor's claim is that, unless one can reduce or 'naturalise' the intentional, then one will be driven to deny that it exists. The motivation for this claim appears to be that, if it is not reducible to something else, nothing as strange and magical as meaning could itself form part of a respectable account of the world. Failing reduction, one would be forced to adopt an anti-realist reconstrual of content as a strictly fallacious way of speaking–a hangover from the pre-scientific age–or as a sort of secondary quality existing only in the eye of the beholder" (Thornton, 1998, 13).

Reductionism's stance on content looks naturalistic since it reflects the conception that mental content needs to be explained as a non-intentional term, viz. naturalizing the intentional properties to the external existing properties of the world. The approach of the Identity theory of philosophy of mind seems closer to reductionism as the identification of the mental states analogously relies on the physical states or the reduction of the mental phenomena supervenes the physical phenomena or brain that opens a metaphysical disjunctive property inherited from the causal-nomological efficacious properties. Wittgenstein laid out an anti-reductionist plea dating back to Brentano (1874) who claims the non-reducible concept of international content in naturalized terms. Besides, the representationalist account that is supposed to expose reductionism describes mental content in terms of the linguistic content that could be given a reductionist elucidation.

The mental states of the speaker enfold an enormous region beyond the internal language system, where it seems intricate to enter into the route of interacting with others. Our critical rationality anticipates that it looks unpromising to explore something as complex as "the communication process". Here we have to segregate and elucidate the related factors that penetrate these complicated actions. Let me articulate one of these factors which is nonetheless the meaning of the speaker's expressions of the internal language. Noam Chomsky once wrote to me, "My mental life is quite different, and I doubt that it is idiosyncratic. I constantly visualize things that I can't express in words, including routes. These are perfectly meaningful to me even if no one else ever has such thoughts and plans, and even if they do, I don't know about them, so they can't determine the meaning to me (unless we accept the externalist doctrines, which, as I've tried to point out, seem to make no sense). In my vague ideas and introspection, there is 'thought' that is inexpressible in language – for example, when I think about how to get to work in the morning, visualizing various routes and options. Or what the weather will be like".[1] Well, one can argue that constructing plans for the day or visualizing the different routes of your work also gets its meaning through language use systems in our society as my thoughts and plans are not only my own but allied to others' thoughts whom we call co-subjects. Chomsky's understanding of meaning[2] will lead to inconsistency. The reason for this is that, in the Chomskyan sense, if you visualize something that is not expressible in words, how could you determine whether the words you use or the visual plans you construct have any truth value or meaning function? For me, the words do not have meaning in the way that shells have pearls naturally. Meanings are not any intrinsic property like shape, weight, etc. of the terms. Words or sentences don't have intrinsic meaning. We (the people) assign or confer meaning to these words. Supporters of meaning internalists like Chomsky or John Searle will urge that a child learns a word each waking hour all through their vital puberty periods of language acquisition. Quite intricate meanings are learned and communicated to others.

Does the child confer the meaning on a particular word? For any internalist, it does not make any sense. They believe in the process of intrinsic interaction of a child's mental structure with a particular word or words. In this process, the external data that triggers the internal tactic parts sound exceedingly narrow. I don't appreciate the opinion and argue against mentalism by saying that "I assume that it is impossible for a learner to grasp the total related body of sentences. In this case, a learner tries to grasp the words of the sentences which are much more relevant to the

[1] I am personally indebted to Noam Chomsky for this valuable discussion.

[2] Looking back to half a century ago, Noam Chomsky sets up the style of his "general linguistic theory" by cutting free from the notion of meaning in his linguistic analysis. He says, "Meaning is notoriously difficult notion to pin down. If it can be shown that meaning and related notions do play a central role in linguistic analysis, its results and conclusions become subject to all of the doubts and obscurities that plague the study of meaning, and a serious blow is struck at the foundations of linguistic theory" (Chomsky, 1955, 141). About a decade later, Jerrold Katz and Jerry Fodor have deductively formalized a model that genuinely nubs of this debate by engaging with a "dictionary view" of word meaning. They said, "It is widely acknowledged and certainly true that one component of a semantic theory of a natural language is a dictionary of that language … Given an accurate dictionary of English, WHICH HE APPLIES BY USING HIS LINGUISTIC ABILITY, the fluent speaker can semantically interpret any sentence on the list under any grammatical derivations" (Katz & Fodor, 1963, 181–182).

knowledge of their expression. So it is not required for the learner to know the meaning of the whole sentence along with the meaning of every word. For me, it is a reliable process of a speaker to know partially the knowledge of the meaning of an expression and later the learning process of knowledge can be gradually increased" (Chakraborty, 2020, 121–122).

Consider the following startling example: we used to believe that whales were fish, but after consulting experts, we learned the truth: whales are mammals. Because scientists first identified the biological characteristics of the whales as mammals that are better expressed externally and socio-linguistically, the conceptual knowledge of "Whales" as fish has been substantially altered in this case. The understanding of the meaning of a sentence or word has two parts. One is a comprehensive grasp of the meaning and another is a partial grasp of the meaning. Ordinary people only understand a portion of what an expert knows, such as that "water" is a liquid that satisfies our thirst, while experts understand the comprehensive meaning of terms like "water", which is the combination of two hydrogen atoms bound to one oxygen atom. However, rather than merely describing the terms, their communication may be made feasible by the referent of the terms that are causally related to them. Numerous instances akin to this exist in which scientific understanding demonstrates an advancement in the definition or meaning of a particular term, but this does presuppose a conceptual shift in the term's or its objective referent. When you adopt the definition, you are indeed introducing a new term into the language system using a reference that we dispose of the old term. Now one may argue whether it is a substantive concern in the model of the theory of content or thought processing. The modification of the definition is not merely a theoretical or descriptional modification that solely depends on the mental content despite being attached to linguistic communication and society. Like Hilary Putnam, one may believe that "The mental event that is someone's thought on a particular occasion is part of the mental life of a subject, and subjective in that sense, but its content (unless it contains the indexical 'I') is shareable".[3]

I appreciate Putnam's thought that successful communication between an idiosyncratic patient and a doctor may be possible if and only if we admit that there is a "Division of linguistic labor" in our society and the doctor as an expert in our society can know the accurate criteria of the syndrome like "arthritis". Besides, our society prescribed that we should accept the doctor's thoughts as they are well aware of the social phenomenon of our language and also know that there is a distinction between conventional phenomenon (where we use our ordinary criteria about an object) and the scientific criterion (a comprehensive analysis of knowledge to know a particular term or object). Even in our society, there is a socially agreed norm that informs us that to identify the specialized criteria or exact nature of an object or term, we should appreciate experts' thoughts. For this social acceptance, successful communication between an idiosyncratic patient and a doctor (expert) may come out. I also think that there is an inclusive knowledge of concept

[3] I am deeply indebted to Hilary Putnam, my mentor, for this valuable note.

possession given by the experts, whereas the common people try to acquire the concept through their own "concept-acquisition" capacities. But any justification of your thoughts or beliefs on "arthritis" or a new word in the place of "arthritis" needs to be verified by socio-lacts and linguistic practices where the "division of linguistic labor" takes a pertinent role (Chakraborty, 2021a, 650–51). Otherwise, our thoughts would be solipsistic. When we learn to know that trees have chlorophyll or trees may process photosynthesis, our concept about trees has been changed, but this conceptual change does not escort any change even in the meaning or the reference of the trees. Can we claim that in this case, the context that gives some new information results in a consequent change in the sentential meaning like "trees are our life"? One can argue that since the expression (Water is H_2O) is nothing but an amalgam of atoms and a technical system invented by scientists, but whatever that object water is, it's not a natural language, so we are incapable of determining the type of the phrase apart from what an agent stipulates.

Chomsky taught us that human language is doubtlessly biological in part (the existence of different languages entails that individual languages could not be completely innate. A similar observation perhaps affects the language faculty case and the process of language learning and creativity of language strongly associated with the universal grammar and the innateness hypothesis. Therefore, the existing changes like introducing new lexical items like "iPhone", "Android", and so on in language are nothing but superficial changes of language that are not the core part of natural language. We manufacture a "camera", but nobody will sensibly consider that we create a new visual system called a "camera". One can also speak on the "new part" of language (as your "surface structure" hints). Still, I think that language is an art of the world that is not created by any individual being. We are just participating in this art of language and following the rules and sometimes contributing something new in the structure of language through the interaction between the subjects and inter-subjects (I, others, plus world). Language goes beyond human creation. We can only practice and express it in our thinking through a propositional form. The ancient natural history of language vindicates that the elementary application of language uses like processing, constructing, evaluating, or especially thinking, etc., all the chief part of the language uses apart from its "reference fixation" rely on the "internal dialogue" of language. On the other hand, the basic concepts are innate in nature, while complex concepts, such as GRAY BOOK, are composed of simpler building blocks. It is simply physically impossible to have them innately hard-wired in the brain.

This leads us to the dilemma of how we understand a term that exists outside of the objective reality and does not fit into our conceptual framework. If this is the case, how may the "potential conceptual scheme" function in human minds? My point is that the conception of the "potential conceptual scheme", by definition, is unfeasible. But we can look at the problem from an "actual conceptual scheme", and in science, there are several cases where the meaning of the term has no interaction with our "actual conceptual scheme", like *curved space-time*, etc. Whenever we have anything recognizable as mental representation, we have something recognizable as mental content, and meaning will be relevant (as one might think of meaning as part of what determines content). Then we can think of meaning without language by thinking of mental representations that are non-linguistic in nature: mental imagery comes to mind. I cannot accept that thinking only is a mental manipulation of representation that does not bear any referential directness to reality! In short, meaning can be discarded from the language in the special case of mental representation that seems non-linguistic. So, the meaning is something that is implanted by humans into our linguistic communication. Besides, getting the meaning of a term without interacting with the conceptual scheme is possible only if our thoughts are located in the domain of mental representation. However, to have a concept for our understanding, it is required to link with the meaning and referent.

In addition, conceptual capacities are, in a sense, brain processes only when the process of their intrinsic parts function as a biological system, but the notion of language use or understanding the meaning of concepts in our communication procedure is based on linguistic ability and socio-linguistic background. The content of a sentence is a function from words to truth values. What a word or a sentence means is an empirical fact, which is true whether one is thinking of individual meaning or the meaning given by a linguistic community. However, I can know what I mean by a word without carrying out an investigation of how my linguistic community uses the word. Most of the knowledge we acquire is based on the procedure of *knowing that* instead of *knowing how*. If a concept is a mental constituent in a thought, where a thought is a truth-evaluable mental representation, then perhaps there is a close connection between concepts and what is expressible in natural language. But even then, there would be cases in which we have to resort to bare demonstratives to capture what a concept is of – consider a concept of a particular shade of a colour for which there is no simple expression in one's language. For me, language is one good way of making public the contents of one's representational mental states, but it is not a precondition to having representational mental states. Noam Chomsky urges:

> The 'referential directness to reality' is using our internal symbolic systems to refer – an action. It doesn't entail that the symbols refer. Take say phonetic representation. The phonetic representation of 'rider' is/rayDr/ (D an alveolar flap). The symbols don't pick out anything in the world, but when our internal computation system uses them as instructions for the articulator organs, a sound wave is produced. Philosophers who have never heeded Wittgenstein's warnings about misinterpreting technical terminology insist that something in the world must be 'represented' by/D/, in their technical sense of 'represent'. Same errors as their technical term 'meaning', 'content', 'rule', etc. That leads them to endless confusion, as in the work you're discussing.[4]

By the way, sometimes our cognitive capabilities entail causal connections in our perceptual schemata. One may compare these theory-forming capabilities to Peirce's abduction.

As far as I can remember, Chomsky sees delimiting the scope of linguistic competence as reducing the method of *knowing how*. He has a need to. I think that here cognitive capacities imposed not only the causal relation in perception, but the

---

[4] My thanks go to Noam Chomsky for this valuable note.

agent's conceptual capacity is directed towards the referential world. So, one's conceptual capacity has some "background enabling status" that maintains its connexion to the object. The slit experiment shows that at some level, the objective world relies on our actions as bystanders. It may well be possible that we can think about numbers, abstract entities, or empty terms, but it is not that the understanding of the terms can only be possible when we attempt to know how to know the terms, we should see that "what the term is", a sort of description that somehow linked to the external world.

Reviving Cultures

Furthermore, one thing that is appealing about the thematic volume *Human Minds and Cultures* is its stand on a striking mechanism of modernizing modern culture, which reconstructs a modern symbolic ideology concerning a cultural transformation towards a theoretical narrative, political hegemony, and technocratic world. There has been a lot of consideration devoted to the Kantian milieu, which opposes any kind of coercion against rational human thoughts and freedom; hence, a modest tuning to the topic of human liberty and cultural equality renews a pragmatic stance to social collectiveness for moving to the welfare of humanity and the socio-linguistic edifice of moral minds (Hauser, 2007). Richard Bernstein wonderfully says, "If we are unwilling to engage in the type of argumentation required for evaluating the conflict of interpretation; if we do not strive to realize the conditions required for practical discourse – then we will surely become less than fully human" (Bernstein, 1978, 236).

An analysing culture paves the way to the amalgam of meaningful actions of the societal members and the representation of the symbolic forms. The concept of cultural understanding is a rubric set forth by several interpretations examining social concern to ensure a range of phenomena together such as freedom, customs, beliefs, convictions, values livelihood, welfare, and assessments of the quality of life. The twofold structures of the culture satisfactorily affirm altitudes of (a) symbolic description and (b) descriptive description.

Basically, to say, what kind of notion the descriptive description is, one has to envisage different arrays of ethnographic connotations adapted to the arena of ethnographic description. Regardless, E. B. Taylor's methodological understanding stands for the descriptive culture that is ambitious to synchronize the whole into the component of parts for entrenching the "scientization of culture" into phenomena range of beliefs, customs, values, laws, and arts, etc. In the scope of the symbolic description, we prioritize the concept of symbols and their interpretation of symbolization. In describing the scopes and limits of culture, I believe, it considerably cultivates the general process of human development that sounds like a requisite part of civilization. Kant's purview of defining culture goes beyond the valued resultant of the cultivation of science and art by putting forth another two pivotal issues – moral laws and refinements that assist humans to become social beings.

The points (descriptive and symbolic) that we nourished here are nothing but the anthropological derivation of cultures, which is nevertheless apt for the scientific understanding and the ethnographic contention of the origin of cultures and so forth. It's exciting when Malinowski revives the concept of cultural variation by excavating into psychological characteristics and cultural adaptations together. He says, "Culture comprises inherited artefacts, goods, technical processes, ideas, habits, and values … Culture is a reality *sui generis* and must be studied as such" (Malinowski, 1931, 621–623). In short, the descriptive culture undertakes a steady expansion of humanity by probing the customs, beliefs, arts, skills, religious practices, weapons, cultivation, etc. throughout the world.

The symbolic culture reflects the dynamic approaches of human beings and non-human beings who produce meaningful expressions in a symbolling form, distinctive criteria. It's relevant to mention here that Geertz urges the concept of culture as "symbolic" (Geertz, 1973, 5), as for him, culture is a *straightforward hierarchy of the meaningful structure* consisting of symbols, dialogues, soliloquies, signs, winks, fake-winks, lampoon, etc. To incept symbolic form, he methodizes the pattern of the culture in terms of cultural analysis and also brings up a radiance of meaning embodied in symbolic forms to construe a communicative interpretation. Culture, as Geertz defines, is a historically transmitted pattern of meanings embodied in symbols, and he also adds later in another passage that a set of control mechanisms – plans, recipes, rules, instructions, – for the sake of governing behaviours (plastic human behaviours basically) that intersects the template for socio-psychological paradigms of organization and institutions as well (Geertz, 1973, 44, 89).

Cultures, for the anthropologists, developed as the assemblage of texts, documents, and social materials, that required interpretation to know about the developments of the cultures and the analogical changes inscribed in implicit as well as explicit ways. Seeing through a structuralist account outlines the internal structural feature of the narratives to undermine the intentional (an intersubjective communication of symbolic forms intended by the subjects), conventional (an interpretation – coined to encode and decode the symbolic forms in retrospect of the subject's convention, rules, and various moral codes, etc.), contextual (embedded in precise socio-historical contexts in spatiotemporal settings) and referential approaches (a pattern of interpreting symbolic forms pinning down something about the referents). In pursuing the development of the symbolic form of cultural values, the embryonic strategies followed by human beings are also copied with their socio-economical function and the reproduction procedure of social contexts, viz. constituting and reconstituting of interpretation over the meaning of the symbolic forms availed in that culture. It looks appealing to pin down a utilitarian approach towards epistemology by saying that knowledge never brings about epistemic justification. An item of knowledge by itself doesn't provide justification for itself, but it could provide justification for some other belief. It's good in so far as it entails more welfare. Similarly, interpretation looks good in so far as it entails meaningful communicative discourse in two different languages to catch up with a new readership who lacks command of the source language. Translation as a humanistic case interprets negotiating between language and cultural asymmetries that opt for meaning in the source text and the implication of the meaning in the target text. It has long been argued that translation goes far beyond meaning and approach to behaviour, cultural mediation, social justice, etc.

As a cluster concept, translation encompasses the *multiplicity of languages* in Von Humboldt's sense, the concept of mind, and multicultural dimensions. The translation should not succeed in any ethnocentric violence or the purview (to translate meant to conquer) that attitude is objectionable even in Nietzsche's words. Spivak talks about translation as a module of social justice for the subaltern and marginalized communities to express their thoughts, livelihoods, and vision of the world with the dominant languages.

In his aphorism on translation, Nietzsche vindicates "Tempo" as the rigid thing to translate from one language to another because, here, the incommensurability of language based on the metabolism of race and culture seems more physiological than syntax and semantics (Nietzsche, 2002, 29). So far, Heidegger has outlined his radical rethinking of hermeneutics, which invokes an enactment of the hermeneutic situation not only as a conceptual scheme but as an objectivity move in any radical interpretation manual. Essence ("what characterizes each 'truth' as truth") and events are primordial parts of the basic ground words, like translating the terminological use of alētheia in Greek as self-event instead of truth in English (Heidegger et al., 2009, 39, 41). Without abandoning the implication of methodological purview on hermeneutics, Heidegger's understanding stems from the ideas underlying the depth of hermeneutics. Hermeneutics, for him, has twofold criteria: First, a philosophical reflection on Dasein, an understanding of the authentic existence of Beings. Second, the methodological reflection on aspects of nature and interpretation as well as several social and cultural concerns.

Philosophers often have a naive view of the processes and practices of translation across diverse cultures over time. Translation deciphers other minds and the cognitive and social functions of individuals and groups in a broad context of global linguistic and cultural variation.

We have a large number of human languages (currently more than 6000), any two of which, in translation, present their dilemmas related to linguistic asymmetries, from dialectics to cultural variation, including literature conventions, religions, moral practices, political ideologies, laws, and customs. Moreover, new variations are always evolving in this philosophical horizon, where often translators have to wrestle with a term that is ambiguous in the source language.

Seeing through the unexpected side effects of translation, let me escalate the twofold functions of language in translation literature:

(a) Language as a map represents the world in more or less accurate ways.
(b) Language helps us to conceptualize what is there in the world.

For me, the primeval position of language in translation brings a cross-cultural avenue, the overarching dimension of the conceptual edifice. On a larger scale, to get an adequate translation of the facts from one language to another, we should not be confined in a culturally specific manner but rather approach a commensurability principle. But in most cases, the linguistic turn conflicts with the curving passage of language and livelihood and opens a path for the incommensurability of translation. So many contents of the world are not describable by our limited capability of thoughts and words, for example, mental images, privileged knowledge, etc. Quine questions whether the translation is ever determinate and reliable. For Quine, translation is a metaphor, analogue, and tool for establishing the larger uncertainties of all communication. Rather, Quine appraises a pragmatic stance consistent with behavioral evidence. In the process of translation, a mode of disparity from one language to another envisages us looking through the distortions and semantic fluxes; it also allows us to register the ambiguities each language carries, their indeterminacy in meaning, their multiculturalism, and their intersection with those of other languages.

After the COVID strike, in particular, the growth of the *Fourth Industrial Revolution*, the luminary economies would settle down knowledge capitalists with distinctly intellectual abilities that cannot be restored by artificial intelligence, and on monetary capitalists possessing artificial intelligence. These superstars will ensue wealth on an excessive scale and devour their wealth to suit their very mixt requirements. This will encourage the surfacing of new service industries. Future technology should be premeditated to allow human beings to live for the future through their dynamic input in the world. In other words, the recent condition in which technology transforms human beings into posthumans with the highest performance for the benefit of the swift escalation of the economy is extolled and must be innovated to bring back the novel liaison between technology, work, and human beings as Being in the world. Instead of playing with hazardous weapons and people's lives, I believe a great world leader will support this approach to propel humanity in new and positive paths. To preserve our humanity and our promise of a better world for coming generations, we must submit to the might of the global forces of humanity. Our modest endeavour would be to usher in the humanistic hymn "on earth, there is one family, and the family is nothing but human" by bringing equality (even in the sense of an equal capacity for cultural adaptation) in the age of diversity and multiculturalism. This thematic volume presents ideas and conceptual models from philosophical, social, legal, and literary perspectives within a truly explicit academic arena, despite the fact that each author had complete autonomy over approach, method, etc. This book includes over a dozen original chapters on the amazing topics that are appended:

1. Human culture from the background of philosophy and scientific affinity.
2. Retrieving the mind in artistic literature and cultural anthropology.
3. A socio-linguistic account of rupturing Minds and Mental contents.
4. Normative outlook on the cultural edifice of the moral mind.
5. Systematizing human minds in the retrospective of multiculturalism.
6. Colonizing life-world and cultural impoverishment in literature.
7. Reviving human minds and cultures in Darwinian and Neo-Darwinian milieus.

In his excellent paper titled ***Thoughts on Morality and Culture***, Jan Narveson vindicates a/the "normative outlook" on "the cultural edifice of the moral mind" to lay out moral affinity and its societal constraints. Narveson's rigorous ploy to define "the moral mind", in actual, nurtures an elementary set of moral principles/values that hold a prescriptive ingress and execute a coherent framework on the whole. Deforming cultural edifice as a human effort and inculcating the notion of moral luck (in the sense of good and bad), the author here shows a plausible stance of affirming moral thoughts. The key question that Narveson handled "Can we override morality" has many facet implications and understandings in our social and cultural paradigm. A comparatively weaker stand that he figures out (like Aristotle, Kant, David Gauthier, and Nicholas Rescher) prioritized a general moral principle that "nobody is to make his way in the world by *imposing costs* on other innocent people".

Harry J. Gensler, in his outstanding paper ***Values and Cultures***, significantly argues against *cultural relativism*, an opinion that denies the notion of objective values in support of good and bad as cultural relatives. Turning to Gensler's contribution, he looks again at *cultural objectivism*, an opinion that expounds the commensurability of objective values with the culture. More specifically, he picked up three objective values that are vivacious and concerned with the relation between humans and across cultures. The first important way of understanding this cultural concord relies on the concept of "Consistency", which pleasurably figured out a claim that humans might be consistent in their thoughts and actions by preserving harmony with moral beliefs and the golden rule in special. The second way vindicates "Knowledge" to evade false belief in the antinomy of our moral beliefs. The third way illuminates "Imagination" which requires an intense and accurate "awareness of what it would be like to be in the place of those affected by our action". Later, he mulls over stirringly different prospects and aspects of evolutionary origins, and their several religious as well as ethical justifications speaking of his distinctive experience of teaching ethics in the United States and China.

Michael Smith, in his startling paper ***Clashes of Culture***, brings up the importance of cultural variation and people's criticism of cultural connotation (relativism) from a different cultural background as per their ways of living. His impressive willingness to retract the doctrine of mutual incomprehension can be demonstrated in a tripeptide structure. The first approach to mutual incomprehension is considered a moral flaw, which might be the prospect of uncomprehending individuals. If only they tried harder, they would come to appreciate the other's point of view. The second approach goes on to intricately describe the shared incomprehension as the product of principled limits on the clarification of someone from another culture. In this paper, Smith dubs the hypothesis that collides with the aforementioned possibility to secure the "culturally mediated ways in which we understand the world are inaccessible to each other". The third possibility, as fostered by Smith, is that mutual incomprehension entailed a metaphysical sphere. If one seeks a companion, then it is easy to make out the contrary of the second possibility, which is reserved for the non-metaphorical sense, in which people from diverse cultures reside in different worlds. The key prospect of the chapter is to analyse these three possibilities owing to rebuff the third.

In his splendid paper ***Hatred: Why Do Such Nice People Do Such Awful Things?*** Michael Ruse revives humans and their social nature – a pattern of self-realization could be evolved when an individual's mind is in the throes of interrelating with a large number of people in society and culture. Another human nature that he looked upon sounds dreadful, which meanders around war and hatred such as the First World War, the Second World War, the Korean War, the Vietnam War, the Russia-Ukraine War, and so forth. This sardonic will is not inert but rather may be seen as an enthusiastic concern towards recuperating the prejudice that the member of one group of the larger communities viciously wishes to rheostat the probable avenues of other communities and their livelihood. Ruse focused on Americans and the vulnerable position of slavery. To pay heed to the paradox, Ruse wonderfully asks, "Why do such nice people do such awful things?" As an evolutionist, he tried to find the answer to the paradox in the periphery of our past history and sociocultural adoption. This rife and contentment of conflict is a module of the ethical failure caused by slipshod human actions and hatred causing great harm to humanity and pushing it into the gulf of oblivion. To persuade this sort of detrimental upshot, one might mull over the aperture of the welfare of humanity, which is nothing but a stratiform evolving towards evocative moral and social living. The very questions that Ruse nourished in the papers are appended below:

(a) "Can we move forward, bringing the knowledge of our biological past combined with the awareness of our cultural present to speak positively and creatively to the challenges that lie before us?"
(b) "Can we moderate or eliminate war and prejudice? The wonderful thing about our human nature is that, although it buckles under the course of history, it does not let that history be the sole determinant".

Deborah C. Poff, in her interesting paper titled ***Neoliberalism, Globalization, and Discrimination in the Twenty-First Century: Issues, Challenges, and Limitations of Habermas' Reason-based Theory of Liberal Democracy in Current Times***, basically delimits Habermas' theory of communicative rationality and deliberative democracy in the context of neoliberalism, irrational anti-intellectual mob behaviour, and globalization. She led to alter the view by highlighting the debacle of a reason-based theory that tactlessly relied on the groundwork of democracy, primarily addressed by outstripping rational and moral understanding.

In her outstanding paper titled ***Legal Emotions: Why Affect Matters for Law***, Antara Haldar relooks at one of the central aspects of law, viz. "internal point of view" introduced by H. L. A. Hart. She contains that pick of her paper nourished on a bridge between jurisprudential literature and advanced moral psychology. The upright of the cognitive sciences takes a germane role in examining what constitutes "the internal point of view". Besides sprouting up the approach, which Haldar called *going beyond* the standard parameter of a cost-benefit tactic in law or economics, and as well as the repeal of morality in the natural law tradition, that undermines the psychological milieu and the innate and encultured based of morality. Following Haidt, Haldar points out that legal systems hold twofold dimensions – either the cognitive (emotional) or the cultural levels, which in actual offers a latent elucidation for why these systems are not only futile to take cause in diversified prospects but also "why they are undergoing a crisis of legitimacy even in places where they are more established"; that looks for a rigorous socio-psychological and empirical experimentation.

Neil Sinhababu, in his stimulating paper ***The Epistemic Argument for Hedonism***, revives moral scepticism in terms of ethical hedonism, a view of expanding pleasure by reducing pain. Having made several differences germane in contrast to spreading a moral agreement, the author speculates enough moral errors to espouse the unreliability of multiplicative moral beliefs. Perhaps a lack of revitalization is possible, like say for the concept of pleasure, which, as Sinhababu picked up, is good through the reliable process of phenomenal introspection, that executes the empirical threshold of moral values in a hedonistic way. An anti-foundational ploy that Sinhababu appraised basically rescinds the singularity of phenomenal introspection as the reliable source of knowing happiness; rather, he bids the ethical hedonism as an intriguing salutation by avoiding unreliable moral sentiment.

In ***The Interpreting Mind and Meaning-Formation: A Relational Critique of Lyotard and Ricœur***, Gerald Cipriani wonderfully portrays the phenomenon of meaning-formation from the Lacanian stance of the jouissance, unconscious, and desire. Although in cultural modernism, meaning-formation deciphers deliverance from conversational order, generating "*jouissance*" of the kind experienced that justifies such an understanding is the *topos* of meaning-formation. Cipriani's concern about the pitfall of the order of discourse that may fall in the realm of the order of *jouissance*, where interpretation comprehends meaning-formation as a non-relational term that evades Ricœur's descriptive hermeneutics of negotiation, a way of fetching diverse methods of interpretation into connection with each other to an assessment between discrete entities. Under the functional aspects of the relational condition that look risky, the author's substantive aspiration goes close to Lyotard's determinant mode of interpretation and Ricœur's hermeneutic methodology to dig out a trenchant understanding of meaning-formation.

A Quinean dictum – "No entity without identity" – is used by Sanjit Chakraborty to explore the philosophical notion of identity of other minds in his paper ***Recurring Dilemmas on Other Minds***. Recent developments on this idea have raised a fresh question: how much do we actually come into contact with the idea of a subjective "mind"? A deeper meaning behind the concept of entity is that some similar entities are distinguished from others by the same characteristic. The author acknowledges that the idea of "other minds" misinterpreted the subjective perspective of privileged access rather than the objective perimeter, which is a reasonable stance. As Chakraborty has pointed out, there is a conundrum that arises when we attempt to summarize these various concerns in the following manner: "Could 'others' be a substance of an individual's mind?" Although this may sound controversial, the author maintains that the other's minds have an adjacent portion. This occurs as a result of the mental content of an individual mind merging in the socio-linguistic periphery, which Chakraborty describes as remaining uncontrollably engrained and tending to function as a conduit between the individual mind and other minds.

References

Bernstein, R. J. (1978). *The restructuring of social and political theory*. University of Pennsylvania.

Chakraborty, S. (2020). *The labyrinth of mind and world: Beyond internalism-externalism*. Routledge.

Chakraborty, S. (2021a). Pursuits of belief: Reflecting on the cessation of belief. *Sophia, 60*(3), 639–654. Springer.

Chakraborty, S. (2021b). Spinning solitude: Coronavirus and the philosopher. *Sophia, 60*(3), 769–774. Springer.

Chomsky, N. (1955). *Semantic considerations in grammar* (Monograph No. 8). Georgetown Monograph Series..

Geertz, C. (1973). *The interpretation of cultures*. Basic Books.

Hauser, M. (2007). *Moral minds*. Little Brown.

Heidegger, M., Gunter, F., & Jerome, V. (Eds.). (2009). *The Heidegger reader*. Indiana University Press.

Katz, J. J., & Fodor, J. A. (1963). The structure of a semantic theory. *Language, 39*(2), 170–210.

Kim, J. (2008). What is 'naturalized epistemology'. In E. Sosa & J. Kim (Eds.), *Epistemology: An anthology*. Blackwell Publishing.

Malinowski, B. (1931). Culture. In *Encyclopaedia of the social sciences* (Vol. 4). MacMillian.

Nietzsche, F. (2002). *Beyond good and evil* (J. Norman, Trans.). Cambridge University Press.

Quine, W. V. (1969). Epistemology naturalized. In *Ontological relativity and other essays*. Columbia University Press.

Taylor, C. (1994). The politics of recognition. In A. Gutmann (Ed.), *Multiculturism: Charles Taylor* (pp. 25–75). Princeton University Press.

Thornton, T. (1998). *Wittgenstein on language and thought: The philosophy of content*. Edinburgh University Press.

Wittgenstein, L. (1953). *Philosophical investigations*. Blackwell.

Wittgenstein, L., & Anthony, K. (1994). *The Wittgenstein reader*. Blackwell Publishing.

Sanjit Chakraborty is an Assistant Professor in the School of Social Sciences and Humanities at Vellore Institute of Technology-AP University. Before that, he taught at the Indian Institute of Science Education and Research Kolkata, the Indian Institute of Management Indore, and the Central University of Hyderabad. His philosophical venture was nourished under the guidance of Professor Hilary Putnam (Emeritus Professor, Harvard University) from 2008 to 2016. His work spans the topics of philosophy of mind, philosophy of language, philosophy of science, ethics and AI and morality. Chakraborty's books include *Engaging Putnam* (De Gruyter, 2022), *Living Without God: A Multicultural Spectrum of Atheism* (Springer, 2022), *The Labyrinth of Mind and World: Beyond Internalism-Externalism* (Routledge, 2020), *Understanding Meaning and World: A Relook on Semantic Externalism* (CSP, 2016), and *Pursuits of Language* (forthcoming). Chakraborty has extensively published many papers in much-respected peer-reviewed international journals, and his works have been reviewed and cited in reputed international journals by noted scholars. Chakraborty has been invited for talks at different renowned institutions and universities overseas. Chakraborty holds the editor and referee positions in the journals of *Human Affairs* (De Gruyter), *Journal of Scholarly Publishing* (University of Toronto Press), *Journal of Experimental & Theoretical Artificial Intelligence* (Taylor and Francis), *Culture and Dialogue* (Brill), *Research Ethics* (Sage), and *Sophia* (Springer) and book reviewer of Springer, Routledge, World Scientific Publishing, and Palgrave Macmillan.

Thoughts on Morality and Culture

Jan Narveson

Abstract Our topic asks for a/the "normative outlook" on "the cultural edifice of the moral mind." In this essay, I shall attempt to fix a fairly definite meaning for each of these notions, and then argue that our normative outlook insofar as this cultural edifice is *moral* is very strong approval. By the (or a) "cultural edifice," I take it, we mean pretty much the whole of society insofar as it is a product of human effort, which in turn is affected, no doubt, by luck – good and bad – and shot through, informed, and constrained by the moralities of the various societies that make it up. But in order to talk coherently of "the moral mind," we need a clear and strong moral theory, one that presents a basic set of moral principles or ideas, ones which ought, therefore, to impose a coherent shape on the whole. This is vague, to be sure, and will be thought in a basic way to be implausible. I shall try at least briefly to affirm the idea and to insist that it is not after all so implausible. Is there, then, a supreme, overriding morality? Yes, I claim (in the wake, to be sure, of many others, including Aristotle, Kant, David Gauthier, and Nicholas Rescher). It has one supreme idea that nobody is to make their way in the world by *imposing costs* on other innocent people. (That formulation is a generalization of the ideas of Kant, Hobbes, Socrates, and perhaps that of mankind generally.) The cultural world as we know it includes the accomplishments of the original and the brilliant, plus even more the communal and cooperative efforts of innumerable people, including symphony orchestras and art museums on and on, down to a gathering of friends playing cricket. On the other hand, of course, it contains much violence and evil. What our world would look like in the absence of that evil is very hard to say – but, surely, it would be *better*.

Keywords Culture · Morality · Social contracts · Liberalism · Kant · Nicholas Rescher

---

J. Narveson (✉)
University of Waterloo, Waterloo, ON, Canada
e-mail: jnarveson@uwaterloo.ca

S. Chakraborty (ed.), *Human Minds and Cultures*,
https://doi.org/10.1007/978-3-031-56448-2_2

Our title might be interpreted as implying that the whole world has but one overall "culture." But of course, that is surely not true. All of us do indeed live in a culture, one or another – but also, those cultures vary hugely from one to another. And it surely will be insisted that it is equally unclear that all of us live under *a* morality, just one – unclear, in short, that we all live in a world united by one single, fundamental morality. Do we? The answer to this last question must be elicited by reasoning – it does not stare us in the face. But I propose to argue that it is so, nonetheless. That's one task to which we will devote attention here – of course, a major one.

Another, surely related, question is, why are we moral (to the extent that we are)? In my view, these are indeed related – though not identical, as I will explain. Well, what is it to be moral, then? Reflection on these questions could take us rather far afield, considering the shelves of books on moral theory that have been written over the centuries. Nevertheless, I hope to show that there is a necessary underlying unity to all of those.

So let's start with the answers:

(1) What is a society's morality? It is the set of requirements that a given society tries to impose on its members. And these, looked at superficially, do indeed vary from one to another. (I talk of "requirements," but there are also the things we praise or blame, recommend, or discourage. For the sake of brevity, I omit discussion of those very important functions.)

But we can understand what a universal morality would be, if there is one: it is the "contract" that *each will rationally make with all others* – thus, the "creed" that is the substance of such a contract. Explaining this will occupy considerable space below, since it is subject to much misunderstanding. I hope to show that, actually, it is pretty obvious. Here one is reminded of the old joke about the two physicists arguing about some arcane thesis, one of whom claims that it is obvious, the other that it is false. They proceed to fill blackboard after blackboard with complicated equations, and after an hour or so of this one of them says, "Ah, you're right – it's obvious!" Well, O.K. We shall see! (But, I promise: no fancy equations!)

(2) We *are* moral (insofar as we are) especially because:

   (a) We were raised that way. If we weren't already, good onlookers stand ready to browbeat us into the right and away from the wrong.
   (b) When we start to think about it, we are glad about (a). We live better lives because we are moral.
   (c) Or we can abstractly contemplate the fact that peaceable, well-behaved societies by and large do well – are good places to live.

(3) Is there, then, a supreme, overriding morality – the "substance," as I put it, of that universal "contract"? Yes, I claim (in the wake, to be sure, of many others, including Aristotle, Kant, David Gauthier, and Nicholas Rescher). It has one supreme idea that nobody is to make their way in the world by imposing costs on other innocent people. (That formulation is a generalization of the ideas of Kant, Hobbes, Socrates, and perhaps that of mankind generally.) Many have

expressed this idea in different ways, to be sure. But I shall in the main stick with this newish formulation.

Yet, might such a morality come unraveled? Certainly it *can*: in the first place, the "morality" we learn at our mother's knee is the morality of those mothers' local cultures, usually, and that culture can be very far off the map. Consider the mothers in Nazi Germany, or of long-ago Sparta. They no doubt inculcated a "morality" – one of total submission to a supreme leader, even though those leaders were, frankly, terrible. It is depressing, and rather surprising, that such leaders have managed to enlist the strong support of so many apparently intelligent (as well as – less surprisingly, perhaps, many more not-so-intelligent) people. How, one must ask, could that happen? We need to understand that, too, for otherwise we have an insoluble problem: why aren't we *always* moral? I go along with almost all in thinking that it obviously just isn't so: some people, alas, are bad. And perhaps all of us, but at least by far most of us, do not always do the right thing. An account of a universal morality needs to explain both why it is what we all *ought* to aim at, and yet also why we often enough fail to do so.

As I say, many thinkers, and many ordinary people, probably deny that there is this one universal morality. Why might people think that? In one sense, the answer is obvious: people suppose that they can gain from a "cultural edifice" that inculcates the values of the conqueror, the acquisitor, the "master." Of course, in so doing, they must either overlook and ignore, or perhaps dismiss as unworthy, the outlook of the conquered, the acquired, and the slave. And in so doing, they must apparently also overlook the potential for people – always the most numerous parts of the societies in question – to unite against those conquerors, etc., turning them, were successful, into exemplars of their own awful intentions. And meanwhile, the wealth of the world will be diminished – often very much diminished. (We read, for example, that seventeenth-century Germany's GDP was reduced by perhaps 2/3 in the course of its Thirty Years' War.)

Why is this fact a "problem"? Roughly, it's that a culture informed by those "principles" will be an awful culture. Its awfulness will usually eventuate in terrible wars, and in the nature of the case, those cultures will in the longer run lose from those wars. It may take a while, of course several centuries, perhaps. Fortunately, it takes much less in recent times: American slavery lasted a couple of centuries and was finally defeated in an extremely murderous war. Nazism, happily, was defeated much more quickly though at a far higher cost (50 million or so dead, including about 20 million of the long-suffering Russians, though only a few hundred thousand of the triumphant Americans.) The real defeat, surely, lies in this. No one but a crank actually thinks there is anything to Nazism anymore, or to the Japanese ideology that animated them during the Second World War. Indeed, both countries are among the foremost liberal, peace-loving countries today. And we can find many examples of really absurd beliefs that have nevertheless been held up as sacred in some culture or other.

To make progress on this, let us remember that our concern is with formulating a *normative* outlook, which means we are far from required to approve of everything that has happened, every culture, etc. – fortunately, since that's scarcely even possible, and certainly not desirable. Those historical cultures have included much that conflicts with many or all others, and so it's impossible to approve of them all. But especially, it's because we are committed to *reason* here – to finding the most rational outlook to recommend to our fellows. Reason cannot recommend the approval of any and all cultures just because they *are* cultures.

Our cultures are where we came from, and so they have the advantage of primogeniture, as it were. It is only when we get out in the world that we come to realize our cultures may not be perfect. (Alas, some people never do learn that.) Not only are they imperfect, but after all, cultures are *made up of* individuals – a whole lot of them. So we have two rival ideas about cultural change: (1) that it is impossible, and every individual in a given culture is imprinted with its various trappings forever; (2) that change does happen, and it is engineered by various particular individuals, or is propelled by them without their realizing it, who bring about those changes – slowly, perhaps even invisibly, but yet, they are real.

The second of these two views is clearly right, and the question is only how it can be – as it assuredly is. The answer, I think, is roughly this. True, we are all brought up in the "same" culture. But, for one thing, the particular people who have the most to do with this – our parents – do differ quite a bit, from one pair (or single parent) to another. Indeed, the differences include differing interpretations of their very culture. And for another thing, we are not all identical (imagine that!): despite, perhaps, the best efforts of those trying to acculturate us, we emerge different – sometimes very different indeed. Our cookie cutters are imperfect, and the very dough from which we are made is not quite the same.

Consider the astonishing variety of individuals in our common culture right now. And then consider how much cultural change there has been in, say, the past hundred years. In making these assertions, of course, I do not pretend to statistical precision, relying instead on commonsense perceptions. But I will be surprised if any think I'm wrong in this. Even in less "individualistic" cultures than America's, cultural variation is amazingly substantial. Does anyone think that Napolean made no difference to French culture? Or Debussy and Ravel? Their work has come close to defining "Frenchness" in music – and yet, each inherited much from their forerunners. The same can be said for the Impressionists in the visual arts. All of those artists were individuals, of course. But also they were all influenced by antecedent culture – again, of course. Yet these people's new contributions were stunning and revolutionary. It happens! And these revolutionary changes lead the way to yet further changes in their wake.

(4) In one sense, morals are *nothing but* the rules of our cultural group. But this obvious-sounding thought is quickly seen to be inadequate – indeed, hopelessly wrong, if it is offered as an analysis of the very idea of morality. Relativism is a theory known to all Philosophy 100 students in the form of "moral relativism." But the fact that morals no doubt have much to do with our culture does not make our culture the supreme authority on what we ought to do. Consider the question of what to do when intercultural conflicts rise – as they inevitably will.

> Problems arise quickly – especially in the case of intercultural activity, where two cultures with conflicting moral teachings intermingle. Now what? The A's aren't going to accept the B's morality – but also vice versa. Can they both be *right?* (Right according to them, yes, of course. But that's not the point. The question is, can they both be right, period? And also, there is the fact that we have long come to realize that cultures are shot through with myth, with what are really antiscientific hunches and worse ... and so on. For such things to infect a morality is for that morality to be disproven. Between science and myth, there is really no contest: science – that is, genuine science, not pseudo-science – wins.

Clearly, a lot of analysis is needed, beginning with an age-old question: which comes first – culture or individual? It's a pretty tricky question, since both answers seem obviously right. No individuals, no culture – of course. And yet we are all shaped, enormously, by our cultures. No cultures, no individuals! How could it not be so? At birth, we are not only helpless but also inchoate. For years, we are immersed in the influences of our cultures, to the point that, some will suppose, we are totally the product of those cultures. But that cannot be either.

Consider the old question – "which came first – the chicken or the egg?" in the light of biological science. It provides the answer: the egg. Once an organism's genome is set, there's no room for further change during its lifetime; but in the process of creating new organisms, things can and do happen. Is there something analogous about cultures? Somehow cultures come to be what they are. It must be that *some* individuals in those cultures *somehow* bring about change – not vice versa. Whatever an individual's culture does to them, it is that individual who can and sometimes does come up with something new that takes hold and spreads. Perhaps in the distant past, some people came to be more influential than others, and their preferences affected the culture and its norms. Perhaps all of culture would be accountable in that way – if only we knew who those people were, what they did, and why others followed their lead. Of course, we certainly do not, and perhaps we cannot have that information. But it's a plausible story, and others, such as the direct effect by the "societal geist," are hopeless when it comes to asking for details, or even clear meaning.

Our analytical project, however, does not depend on the historical details. What is needed is explained well enough by game theory. Our proposition is that we are all better off when our society is guided by certain very strong and very general rules, the basic one of which is that no one is to deliberately inflict costs, damage, or hurt, on anyone else, and this regardless of which tribes the participants may belong to. Why think we might all gain from this? Because we are each thereby given the security to pursue our lives as we see fit – and, because if we think we can take advantage of some weaker or unsuspecting person on a given occasion, others can think the same, and then the advantage disappears – indeed, it turns into a disadvantage – and we all lose by having to be part of the "policing" regime to try to prevent that.

Too many have objected to the Social Contract view of the fundamental foundations of morality by insisting that the principle of keeping your agreements is itself one of the principles in need of foundation – no? Answer: No. As Hume himself saw,[1] the "agreement to keep our agreements" is not one of the several principles agreed to in the Social Contract. The analysis of the latter is game-theoretic. Its foundation is, simply, Reason. What it is, is the right set of Rules For the Group (as applied to the "group" of *everyone*). What they say to everyone is, "Whatever you do, be sure it's compatible with these rules!" This gives everyone, in the words of (early) Rawls, the maximum freedom compatible with the extension of a like freedom for all. That's why we "sign": each of us gets the best "deal" we can, given what all the others are like.

At the core of this is the Non-Harm Principle. *Whatever* we do to our fellows, we do *not* take over their lives, or any part of their lives, for our own purposes without *their* consent. This establishes the basic relationship between each of us and each of the others. We are, each of us, persons – individuals – with our own lives to lead. Anything anyone else does to us is either going to contribute positively to our lives as we live them or at least not "contribute" negatively. Our interactions with others should either leave us where we are or get us more or less ahead. If what we all know is that this is what we can expect from our fellows, then we have the excellent general basis for getting along.

Of course – as we must immediately point out – this "security" we are "given" is in practice sometimes one we don't get – in too many particular cases (it's worse, too, for persons of differing color from the majority). And even when we do get it – as when our cultural mores are on "our side," as they usually are, more or less – there is the possibility that those mores will also be mixed with a good deal of plain nonsense, and not a little in the way of what economists call "rent-seeking." How much does that matter? That depends on just what we are asking. Any actual program or initiative in the real world will by and large be fraught with special considerations. When we are just theorizing, we can factor such things out, but in the "real world," they are, often (though wrongly) seen as the very stuff of the matter. Reason supports the Social Contract. That is not to be taken lightly – for example, morality is not to be subsumed by a parochial concern such as the symbolic apparatus of one's own religion. Social Contract always looks to the universal. But the people who must deliver its basics often enough have other baggage besides that basic thing. We may not have too much choice about it – that is, about which is the "basic thing" and which is the "baggage" in those particular circumstances. We make our way through our lives as best we can.

It is in this, I think, that there lies the source of pessimistic views about morals and culture. For example, virtually, all religions are highly involved with codes of morals, and that perhaps leads some people to think that religion *is* morality. That one who thinks that way has it wrong is obvious when we just ask: well, what about

[1] For a discussion of Hume's signal contributions, see "The Agreement to Keep Our Agreements: Hume, Prichard, and Searle" (Narveson, 1994) For a more general argument, see "The Relevance of Decision and Game Theory to Ethical Theory" (Narveson, 2010: 497–520).

the proponents of any of the other several *thousand* religions, then? Each of them thinks that their religion has the truth of the matter. Of course, a minimum of all but one of all those religions must be wrong, right? And how is it that it just happens to be *one's own?* Even if there weren't a basic logical case against the thesis – as discovered by Plato over two thousand years ago – this fact would surely be enough to give the would-be rational proponent of religious morals pause.

In brief, what we need is Liberalism. "'Whatever happened to good old liberalism?' Pinker said to me later, exasperated but cheerful. "Who's going to actually step in and defend the idea that incremental improvements fed by knowledge, fed by expanding equality, fed by liberal democracy, are a good thing? Where are the demonstrations, where are the people pumping their fists for liberal democracy? Who's going to actually say something good about it?" [from Alex Blasdel, "Pinker's Progress – the celebrity scientist at the centre of the culture wars" – Guardian, Tue 28 Sep 2021] Well – I am, for one. Now, Liberalism is not the same as liberal *democracy*. Liberalism is the general view that politics and morals are to serve the end of enabling each individual to live their *best life*, as seen by that individual. To enable each to do this, we of course run into the problem that people's best lives, as seen by them, can be impossible to coexist. And when they do, what do we do? That's where liberal *morality* kicks in. In principle, we may do as we like, *so long as* what we like does not impose on others. That may be hard; that formula will still involve us in problems. Interpersonal conflict is not so easily avoided. But it can succumb to careful analysis.

So in practice, this is where we bring in Democracy. The democratic system says that in the face of inevitable conflict, we put issues to a vote. More precisely, we nowadays put candidates, who hopefully share our points of view, to a vote. The connection to particular voters thus becomes very fuzzy. And on top of it, it's far from obvious that the outcomes of democratic elections will always be better than they could have been – as practically every thinking citizen will likely agree. So why democracy? Because, in brief, as the old saying (of Winston Churchill and numerous others has it) every other among the familiar alternatives is worse, and most are *much* worse. And politics, again, interacts very closely with culture. Culture affects politics and vice versa.

Which brings us back to the general subject of Culture. Any particular culture will be shot through with values. And what makes a culture the culture it is, is the fact that many or most of those values will be, quite widely, *shared.* While there will always be much about our cultural values that we do not know, or do not understand, there will also be much that feels natural and obvious – as is confirmed when we encounter a stranger, who quite obviously is not familiar with how "we" do things around here. Yet however strange their ways, we have a pretty good idea when we would have gone too far in our dealings with them. At the core of every culture, there is one rule: whatever you do to or with this person, you are not to fundamentally hurt or harm them. Always, the basic idea is that their lives are *theirs* – not anybody else's – and that as such, they are to be permitted to live those lives, insofar as that can be done consistently with the like freedom of others. (Consistency among the different elements of their own values is another, and very difficult, matter. Again, we high-handedly assume such consistency in this short discussion.)

Culture can too easily recast this essentially individualistic doctrine so as to foist it on all identifications with various groups – ethnic, sexual, or what-you-will. Is it a Western bias to insist on resisting these? No. After all, if we allow that, people may choose their identifications, thus allowing the possibility that virtually all in a certain group will have chosen their membership in that group, we capture a large part of what the "Eastern" view wants. But those few on the fringe who refuse to go along? The "Western" view says they are to have their way. And it is difficult to deny that this "Western" view basically has it right.

Why? My answer – perhaps due in part to my lifelong status as a "Westerner" – is that the "Western" view is. There is at least one very strong reason for this: the Social Contract – the "agreement" of all with all that constitutes the world's basic morality – needs to benefit *all* of its "signatories" – who are all of us. Only in that way can we ensure a genuinely universal "subscription." As I have argued, that is surely the only way to guarantee that result. (Narveson, 2016: 695–711).

But that leaves a huge amount of room for human diversity, which we can all enjoy, both as members of the particular culture(s) we belong to and as spectators of many of the others. I don't think we can do better than that. Should we *want* to?

Those who insist that the Western view is too "thin" should note that they are insisting on the *rights of groups to exercise coercion over the individuals within it.* Then the question must be, where did this right come from? Suppose G is a group into which you are born – no choice there. But G has practices from which member A dissents. Why would G – that is to say, those who claim to be arbiters of G's view or outlook – be entitled to prevent A from expressing or even privately engaging in that dissent? It is unhappily typical for groups to do that. In India and many other places, for example, women are often killed for insisting on choosing their own husbands instead of submitting to group rule.

Indeed, such practices are, in one form or another, engaged in throughout the world. That they are raises the question, in stark form: Should we side with individuals or groups? The individualist answer, as we might call it, is pretty evident: with the individual. But a simple assertion on a matter like that is surely unsatisfactory. Anybody can assert – but where is the argument? It seems a matter of simple metaphysics: individuals are "prior" to groups in the straightforward sense that groups *consist* of individuals and, importantly, not vice versa. Only individuals can address and discuss such questions.

Some will no doubt complain that to imagine an individual persisting through time in the absence of any sort of group contact is to imagine a nobody, a nullity. And of course, that's true. Recall, for example, the absolute dependence of infants on parents: there is no way that an individual can assert himself there, as a completely separate entity.

Yet, even in that context, there must be the physical body of the child, identifiable as such, in order even for these relations of dependency to be instantiated. (And as any parent can attest, infants can make their occasional dissent very loud and clear!)

We must, of course, distinguish between the mere physical object that is our body, and the person we grow up as – manifesting personality, character, and intellect. It is in these latter respects that "culture" becomes relevant. And as I have agreed, the huge influence of our cultures – very broadly speaking – on each of us is beyond doubt. And yet, we differ. Our individual quirks, special abilities, and – perhaps most important of all – our individually thought-out ideas survive the influence of culture.

To be sure, "survive" is much too weak a word. For we may conjecture that as time goes by it is the individual contributions of each person that go to make up the stuff of culture – its warp and woof. Note that these individual contributions aren't necessarily for the good. Some are, some aren't. The evil persons among us too often have too much influence. But that's the point: both good and bad testify to our abilities to shape our cultures just as they shape us.

So, why – asks our Rational Individual – should I stand with the "good guys," the "nice," the respecter of others' persons? One good answer – perhaps the best – is that I have most to gain in a community that has adopted and internalized the Hobbesian First Law of Nature, together with its implications. The security afforded by the prohibition of harm to others, just because that benefits the agent, is expectedly much greater than in a community of thugs and criminals (who, of course, would have little to gain among people who'd think nothing of killing them for their money.) Beyond that lies the string of benefits that Hobbes lists: society, learning, and the rest of it.

Do we live in such a society? Pretty close. Most of us know that our neighbors aren't about to rob or murder us. Much farther down the street, perhaps – and we willingly support police to make life much shorter and more dangerous for the criminal types. And beyond that lie the sciences, the complex literatures, the inspiring musics, the beautiful people with their high fashions (which we of modest means can cheaply enjoy at a distance), the nice, comfortable, efficient automobiles, and so much more. These are, I suppose, the values of bourgeoisie. And indeed, they are. (McCloskey, 2006, 2019) Luckily for us!

The moral mind, then, is the key to all this. Don't knock it!

References

Narveson, J. (1994). The agreement to keep our agreements: Hume, Prichard, and Searle. *Philosophical Papers, 23*(2), 75–87.

Narveson, J. (2010). The relevance of decision and game theory to ethical theory. *Ethical Theory and Moral Practice, 13*(5), 497–520.

Narveson, J. (2016). Social contract: The only game in town. *Dialogue, 55*(4), 695–711.

McCloskey, D. N. (2006). *The bourgeois virtues: Ethics for an age of commerce*. University of Chicago Press.

McCloskey, D. N. (2019). *Why liberalism works: How true liberal values produce a freer, more equal, prosperous world for all*. Yale University Press.

Jan Narveson is Distinguished Professor Emeritus of the University of Waterloo in Canada. He is the veteran of some hundreds of presentations and colloquia in North and South America and Europe, and author of seven books, notably *The Libertarian Idea* (1988 and 2001), *You and the State* (Rowman & Littlefield, 2009), and *This Is Ethical Theory* (Open Court, 2010); plus several hundred articles and reviews in philosophical journals and collections. He has also been active in the presentation of classical music concerts (some 2500 so far) as president of the Kitchener-Waterloo Chamber Music Society. He was elected a Fellow of the Royal Society of Canada in 1989; in 2003 he was made an Officer of the *Order of Canada*, which is that country's highest recognition of civilian achievement.

Values and Cultures

Harry J. Gensler

Abstract "Values and Cultures" argues against *cultural relativism* (which denies objective values and holds that good and bad are relative to culture) and argues for *cultural objectivism* (which holds that cultures tend to share a common core of objective values). I begin by trying to make a plausible case for cultural relativism; then I point out problems with this view. I argue for three objective values that are widely shared across cultures. **Consistency** claims that we ought to be consistent in thought and action; this includes things like consistency in beliefs, ends and means, living in harmony with our moral beliefs, evaluating similar cases similarly, and treating others only as we're willing to be treated in the same situation (the golden rule). **Knowledge** claims that we need to be informed and that false beliefs are to be avoided. **Imagination** claims that we need a vivid and accurate awareness of what it would be like to be in the place of those affected by our actions. Further sections deal with our evolutionary origins, various metaethical justifications, religion and ethics, and my experience of teaching ethics in the United States and China.

Keywords Cultures · Values · Cultural relativism · Moral realism · Golden rule

"Values and Cultures" is a big topic and can be approached from many angles. Here I'll contrast two views. The first I call "**cultural relativism**," which holds that *good* and *bad* are relative to culture (so there are no objective values). The second view (for want of a better term[1]) I call "**cultural objectivism**," which holds that cultures tend to share a common core of objective values. I'll argue for cultural *objectivism*

Harry J. Gensler died before publication of this work was completed.

[1] "Cultural objectivism" is a strange term (I just invented it), but it's useful to have this named alternative to cultural relativism. While *cultural relativism* is a specific metaethical view (about the meaning and objective status of ethical terms), *cultural objectivism* is compatible with various metaethical views (which may see the common core of objective values as based on things like self-evident intuitive truths, divine commands, what an ideal thinker would favor, universalizable prescriptions, and so on). We'll discuss this later.

H. J. Gensler (✉)
Loyola University, Chicago, IL, USA

S. Chakraborty (ed.), *Human Minds and Cultures*,
https://doi.org/10.1007/978-3-031-56448-2_3

over cultural *relativism* (CR). And I'll clarify these views further and give arguments for or against them.

First, I'll defend cultural relativism (CR) as strongly as I can. In section "Pretending to defend cultural relativism", to make this more vivid, I'll pretend to hold CR myself. Later sections will point out problems with CR and give what I see as a better approach.[2]

Pretending to Defend Cultural Relativism

As we grow up, we tend to take on the values of our culture and see these as objective facts. When I was a child, my culture taught me things like "Infanticide is wrong" (a value judgment) and "Warming ice cubes melt them" (an empirical judgment); at first, I took both to be objective facts (plain truths). Later, I saw that there are many different cultures, often with very different values. Attitudes vary with time and place; morality is a cultural construct. The values that I was taught were the values of my own society—not objective values that all cultures agree on. Consider my two examples.

1. "Infanticide is wrong" is a *value judgment*; it's true in my culture but not true in some other cultures (like ancient Rome, where killing babies was common).
2. "Warming ice cubes melts them" is an *empirical judgment*; it's true in all cultures (or at least in all that live where some water gets below freezing).

CR's core idea is that evaluative terms (like "good," "bad," "right," "wrong," "ought," "virtuous," "vicious," "fair," "unfair," and so on) are *relative terms*; they always need a reference (sometimes implicit) to a particular culture to complete the meaning. So things aren't "good" or "bad" objectively (independently of culture); rather, they may be good in one culture but bad in another.

"On the left" is another example of a relative term. Suppose I stand in front of the classroom, with the door on my left. I ask my students, who face me, "Is the door on the left or on the right?" They answer, "The door is on *your* left but on *our* right"—or perhaps "The door is on the left *relative to you* but on the right *relative to us*." I then ask, "But in an objective way, not relative to perspective, is the door on the left or on the right?" They answer, "It makes no sense to ask this, since 'on the left' and 'on the right' are relative terms." I answer, "Yes, and that's how we cultural relativists see 'good' and 'bad'; things are good and bad only in reference to a given culture, and nothing is good or bad objectively."

---

[2] My treatment of cultural relativism is an expanded version of Chapter 1 of my *Ethics: A Contemporary Introduction*, 3rd ed. (New York: Routledge, 2018); I'll later refer to this book as "*Ethics*." I'll refer just by title to these further books of mine: *Formal Ethics* (New York: Routledge, 1996), *Ethics and the Golden Rule* (New York: Routledge, 2013), *Ethics and Religion* (New York: Cambridge, 2016), and *Introduction to Logic*, 3rd ed. (New York: Routledge, 2017).

Some claim that things *are* objectively good or bad. These people are merely objectivizing the values that their society taught them. They think, for example, that infanticide is *objectively* bad—and they think this because their society taught them this; had they been brought up differently, they likely would have different beliefs. So objectivists commit the error of seeing the attitudes that their society taught them as objective values. Our society teaches us merely its own values, not what's objectively good or bad, and "objectively good or bad" makes no sense.

Objectivists have a "We're right and they're wrong" intolerance toward conflicting values of other cultures. In contrast, we cultural relativists think that conflicting values of another culture aren't *wrong*, but only *different*. We try to understand the values of other cultures in terms of *their* beliefs and values; we don't just reject their values because they clash with ours. Cultural relativism thus leads to tolerance and understanding toward other cultures.

Here's another example, based on the Wonder Woman stories[3] and on American history.[4] Consider these three cultures:

1. Diana, later to become Wonder Woman, was born on the island of Themyscira. This is the home of the Amazons, a tribe of women warriors. Themyscira is a monarchy ruled by a queen. In the Amazon culture, only women ought to rule.
2. In early America (before roughly 1920), no women ought to vote or hold office.
3. In recent America (since roughly 1920), men and women have an equal right to vote and hold office.

Here three cultures have three different attitudes about the role of women in government. We cultural relativists wouldn't ask "Which attitude is *right*?"—at least if this means "Which attitude is *objectively right*?" Attitude are right or wrong, not objectively, but only relative to culture. There's a strong inclination to think that the values of *our* society are the objectively right ones, but this must be resisted.

The *myth of objectivity* says that things can be good or bad "absolutely"—not relative to this or that culture. I have three arguments for rejecting this and moving to cultural relativism.

1. My *cultural differences* argument points out that cultures can differ radically on moral issues, like infanticide, polygamy, and women's rights. When we speak of good or bad *absolutely*, we just absolutize our society's norms and take them to be objective facts; so, in dealing with another culture's conflicting norms, we think that *we're right and they're wrong*. Believing in objective values is provincial and narrow-minded; those who accept this myth of objectivity need to

[3] Wonder Woman is a fictional comic book superhero created in the 1940s; see John Sazaklis and Luciano Vecchio (2015). See also two recent movies: *Wonder Woman* (2017) and *Wonder Woman 1984* (2020).

[4] In the United States, the 19th amendment to the Constitution (ratified on August 18, 1920) granted women the legal right to vote. The related 15th amendment (ratified on February 3, 1870) granted men (male adults) of every race the right to vote. Both amendments took much time to be put into practice in some areas.

discover the diversity of values by studying anthropology or living for a time in another culture.
2. My *product of culture* argument begins by seeing that societies create value systems and teach them to their members. We shudder at the idea of infanticide, because we were taught to shudder; had we been brought up in ancient Rome, we'd see infanticide as perfectly normal and fine. Societies create different styles of clothing, different types of food, different ways of speaking, and different values. The clothing styles and the values of another culture aren't objectively right or wrong; they're just different. We need to recognize the deeply cultural basis for morality.
3. My *no-neutral standpoint* argument points out that there's no neutral rational standpoint for criticizing another culture's values. Experiments can decide scientific issues; if someone thinks that heavy objects fall faster than light ones, we can drop objects of different weights and see which hits the ground first. Moral issues aren't like this. When we argue about ethics, we just assume the values of our own culture. There's nothing objective here.

As I've come to believe in cultural relativism, I've grown in my acceptance of other cultures. I was once a foreign exchange student. In the beginning, I had this "we're right and they're wrong" attitude. I struggled against this. Later I realized that the other side isn't "wrong" but just "different." We have to see others from their point of view; if we criticize them, we're just imposing the standards of our society. We cultural relativists are more tolerant.

Through cultural relativism, I've come to be more accepting of the norms of my own society. CR gives a basis for a common morality within a culture—a basis that pools everyone's ideas and ensures that norms have wide support. So I can feel solidarity with my own people, even though I respect other groups with different values.

Problems with Cultural Relativism

The core of **cultural relativism** (CR) is that "good" means "socially approved." As a definition, it says that "X is good" means "The majority (of the society in question) approves of X." Unless otherwise specified, the society is that of the person making the judgment; when I say "Hitler acted *wrongly*," I mean "according to the standards of *my* society." By CR, we pick our moral principles by following what our society approves of.

While CR at first seems very plausible to many people (including college freshmen), a deeper reflection reveals many problems with CR and shoots it full of holes.

One of CR's big problems is that it forces us to conform to society's norms—or else we contradict ourselves. If "good" and "socially approved" mean the same thing, then whatever is one has to be the other. So this reasoning would be valid, and we could prove that something is *good* from the premise that it's *socially approved*:

X is *socially approved.*

∴ X is *good.*

And this statement would be self-contradictory:

X is *socially approved* but isn't *good.*

If CR is true, then we have to conform completely to our society's values—we can't consistently disagree with them—we aren't free to think for ourselves on moral issues. This is an absurd result. We surely *can* consistently disagree with our society's values. We can consistently affirm that something is *socially approved* but deny that it's *good.* This would be impossible if CR were true.

If you're a convinced cultural relativist, you could try to accept these seemingly implausible results; you could hold that it *is* self-contradictory to disagree morally with the majority. But this would be difficult to hold consistently. You'd have to hold that civil rights leaders contradicted themselves when they disagreed with accepted views on segregation. You'd have to conform to the majority view on all moral issues—even if the majority is ignorant. And if majority opinions change, you'd have to change your moral beliefs too. With CR, the central virtue of the moral life is *conformity* (being a follower instead of an independent thinker); *good actions* are ones that are *socially approved.* By outlawing disagreements, CR would stagnate society and violate the critical spirit that characterizes philosophy.

If we apply CR to concrete issues, we get crazy results. I'll give five examples.

1. *Racism.* Imagine that you live in a society that practices and approves of extreme racism—perhaps America before the Civil War (black slavery) or Nazi Germany (killing Jews). A satisfying view of morality should show how to attack racist actions; CR fails at this, since it holds that racist actions are good in a society if they're socially approved. Even worse, CR logically entails that protesters who say "Racist actions are *socially approved* but not *good*" contradict themselves. These CR implications are difficult to accept.
2. *Global warming*. Roughly speaking, there are two main views. *Climate change affirmers* say the earth is rapidly warming, this is mostly caused by human activity, and humanity ought to make radical changes, especially in its use of fossil fuels, to prevent catastrophic harm for future generations. *Climate change deniers*, in contrast, claim that human activity isn't a major cause for recent temperature increases (which happen for other random causes), and humanity needn't change its use of fossil fuels. If we followed CR consistently, we'd go with whatever view was *socially accepted* by the majority; this is the "good" view, even if people accepted it from ignorance of scientific evidence. Applying CR to global warming could bring disastrous consequences to humanity.
3. *Teaching morality to our children.* If we accepted CR, how would we bring up our children to think about morality? We'd teach them to think and live by current social norms—whatever these are. We'd teach conformity instead of critical thinking. We'd teach that these are correct reasoning: "My society approves of A, so A is good," "My peer group approves of driving while drunk, so this is good," and "My Nazi society approves of racism, so racism is good." Our children will

grow up to be conformists who think the "socially accepted" way is always the "good" way. Applying CR to moral education would have sad consequences.
4. *Tolerance toward other cultures*. Whether such tolerance is good depends on the attitudes of your society. If your society approves of making fun of people of other cultures, hurting them, and discriminating against them—then these are good. CR can lead to extreme intolerance toward other cultures, if your society approves of such intolerance. This removes one of CR's biggest attractions.[5]
5. *Independent moral thinking*. Whether this is good depends on your society's attitudes. Suppose your society approves of adults thinking for themselves on moral issues and sometimes disagreeing with accepted norms. Then on CR, such independent moral thinking is good, but if you do it then you contradict yourself—since it's self-contradictory to say "Such and such is socially accepted, but it isn't good." So CR here leads to incoherent results.

CR may sound good when viewed abstractly, but CR applies poorly to issues like racism, global warming, teaching morality to children, tolerance toward other cultures, and independent moral thinking.[6] And it's really disappointing if you want to reject the status quo and make the world a better place.

---

[5] It grounds tolerance better if we hold that it's *objectively true* that, in general, we ought to respect and allow actions, beliefs, and attitudes of other individuals and cultures, and dissenting views in our own culture, even when these differ from our own views. "In general" permits exceptions. We shouldn't respect or allow hateful Nazi racist actions. And we should correct our children when they say "I hate people of such and such a group—because they're all …" (giving a false stereotype). But in general, tolerance should rule.

[6] Cultural relativists often tend to think like moral realists when they evaluate the norms of their own society. Here's an example. The anthropologist Ruth Benedict defended CR ("A defense of cultural relativism," *Journal of General Psychology* 10 (1934), pp. 59–82): "We recognize that morality differs in every society and is a convenient term for socially approved habits. Mankind has always preferred to say, 'It is morally good,' rather than 'It is habitual' …. But historically the two phrases are synonymous." But when writing about racism later (Mead, 1959: 367–8, 359–60), she ignored her CR and saw racism as both *socially approved* and *bad* (like a sickness). She used anthropology to counter ideas about racial inferiority, claiming that the top third of the human race (physically, intellectually, or morally) includes all races. She urged America to "treat people on their merits, without reference to any label of race or religion or country of origin" and "ensure human dignity to all Americans"—even though these ideas clashed with her view that good = socially approved. While CR may seem sophisticated when we study other cultures, CR gives a crude way to actually reason about moral issues.

Cultural Differences

Moral realism claims that *some* things are objectively right or wrong, independently of what anyone may think or feel.[7] For example, Dr. Martin Luther King[8] (1963) claimed that racist actions were objectively wrong. Racism's wrongness was a fact; any person or culture that accepted racism was mistaken. King didn't *absolutize* accepted norms; instead, he *rejected* them. He appealed to a higher truth about right and wrong, one that didn't depend on human thinking or feeling. He appealed to objective values.

Cultural relativists reject this belief in objective values and call it "the myth of objectivity." In their view, actions are good or bad only relative to this or that culture; against King, actions can never be good or bad objectively. But are objective values really a "myth"? Against objective values, we gave the *cultural differences*, *product of culture*, and *no neutral standpoint* arguments. This is the first argument:

> (1) My *cultural differences* argument points out that cultures can differ radically on moral issues, like infanticide, polygamy, and women's rights. When we speak of good or bad *absolutely*, we just absolutize our society's norms and take them to be objective facts; so, in dealing with another culture's conflicting norms, we think that *we're right and they're wrong*. Believing in objective values is provincial and narrow-minded; those who accept this myth of objectivity need to discover the diversity of values by studying anthropology or living for a time in another culture.

Let's express this reasoning more clearly. The desired conclusion is "No moral beliefs are objectively true." Here's a strict premise-conclusion formulation:

No idea on which there is wide disagreement is objectively true.
All moral beliefs are ideas on which there is wide disagreement.
∴ No moral beliefs are objectively true.

While the conclusion follows validly from the premises, both premises are false or doubtful. Against premise 1, a wide disagreement doesn't show that there's no truth in the matter, that neither side is right or wrong. There's much disagreement about cholesterol, religion, and the causes of global warming; yet there may still be a truth of the matter here.

Premise 2 is also false or doubtful. While some moral beliefs have wide disagreement, others have wide global agreement.[9] Most cultures have somewhat similar norms against stealing, lying, adultery, and killing. Many moral differences come from applying similar basic values to different situations. The golden rule (GR), "Treat others as you want to be treated," is widely accepted across cultures. And the world religions and the United Nations have consensus statements about ethics. Later I'll argue that cultures tend to share a common core of objective values.

---

[7] What I've called "cultural objectivism" accepts moral realism but adds a factual claim, which most moral realists would accept, that cultures tend to share a common core of objective values.

[8] "Letter from the Birmingham jail," available in many places on the Web (search for the title), for example https://kinginstitute.stanford.edu/sites/mlk/files/letterfrombirmingham_wwcw_0.pdf

[9] For a summary of scientific research about ethical universals, see Kinnier et al. (2000: 4–16).

Cultural relativists tend to distort ethical differences in at least six other ways.

1. CR sees the world as neatly divided into *morally uniform societies*. While there are big moral differences *between societies*, there's little moral difference *within each society*, since the majority decides the norms. But the world isn't like that. Instead, the world is a confusing mixture of overlapping societies and groups; individuals in a given society, far from simply following the majority view, may differ widely among themselves on moral issues. Consider speed limits. We pretty much agree on what the "official speed limit" for a given place is (since the government decides this), but we disagree on whether it's *morally proper* to go much over the speed limit (since this expresses our individual beliefs).

CR leads us to see other societies in oversimplified stereotypes. Some people say, "We Americans respect human rights and oppose terrorism, but society X believes in terrorism and opposes human rights." But both societies likely have many people on both sides of the human rights issue. In America, the news often has a story about white supremacists who act violently against blacks or Muslims. And society X likely has *some* individuals who push for human rights and fair treatment for all. CR encourages a **we** versus **they** mentality, where both groups are looked upon in oversimplified ways.

2. CR ignores the *subgroup problem*. We all belong to overlapping groups. I'm part of a specific nation, state, city, and neighborhood, and I'm also part of family, professional, religious, and peer groups. These groups often have conflicting values. According to CR, when I say "Racism is wrong," I mean "*My society* disapproves of racism." But *which* society does this refer to? Maybe most in my national and religious societies disapprove of racism, while most in my professional and family societies approve of it. CR could give us clear guidance only if we belonged to just one society. But the world is more complicated than that. We're all multicultural to some extent.
3. CR says that those who believe in objective values just absolutize values they learned from their society. This needn't be so, as we see from Martin Luther King. In holding that racism is objectively wrong, he *disagreed* with accepted norms—he didn't *absolutize* them.
4. Moral realists needn't say "We're right and they're wrong." Perhaps there's a truth to be found in moral matters, but no culture has a monopoly on this truth. Different cultures may need to learn from each other. To see the errors and blind spots in our values, we may need to see how other cultures do things, and how they react to what we do. Learning about other cultures can help to correct our cultural biases and move closer to the truth about how we ought to live.

Cultural relativists rightly reject the dogmatic "We're right and they're wrong" attitude. And they stress the need to understand the other side from their point of view. These are positive ideas. But their "Neither side can be wrong" limits our ability to learn from other cultures. My side might be wrong (e.g., about gun control); learning about other societies (and how they limit access to guns to reduce killings) may help my side to correct its mistakes. On specific issues, moral realists may say "They're right and we're wrong."

5. Moral realism can respect *most* cultural differences, even though it insists that *some* things are objectively right or wrong. Different societies often have different ways to promote common objective values. Traffic laws everywhere ought to promote traffic safety, but societies legitimately differ in which laws they set up to do this (for example, whether we all drive on the left or all drive on the right). And all societies ought to have rules and customs that promote the goods of family life (like loving relationships and raising children), but societies differ on how to do this most effectively for their people (perhaps with arranged marriages in some places). Many cultural differences are *experiments in living* that test different ways to promote common values (like healthcare for all at an affordable price). Often cultures can learn from each other; so if we have a problem with weak family life or expensive healthcare, we might study other cultures who handle these areas better.

Moral realism can accept strong human rights principles. It can hold that hurting people because of their race is *objectively wrong*—and that any society that approves of this needs to change. CR cannot accept such things. Respecting most cultural differences doesn't make you a cultural relativist. What makes you a cultural relativist is the claim that *anything* that's socially approved must thereby be good (including hurting people because of their race).

6. CR does little to establish common norms *between* societies. As technology shrinks the planet, moral disputes between societies become more important. Nation A approves of equal rights for women (or for other races or religions), but nation B disapproves. What is a multinational corporation that works in both societies to do? Or societies A and B have value conflicts that lead to war. Since CR helps very little with such problems, it gives a poor basis for life in the twenty-first century. Maybe moral realism can do better.

Product of Culture

Here's our second argument—and a stricter formulation:

> (2) My *product of culture* argument begins by seeing that societies create value systems and teach them to their members. We shudder at the idea of infanticide, because we were taught to shudder; had we been brought up in ancient Rome, we'd see infanticide as perfectly normal and fine. Societies create different styles of clothing, different types of food, different ways of speaking, and different values. The clothing styles and the values of another culture aren't objectively right or wrong; they're just different. We need to recognize the deeply cultural basis for morality.

No products of culture express objective truths.
All moral beliefs are products of culture.
∴ No moral beliefs express objective truths.

Again, both premises are false or doubtful. Against premise 1, some products of culture express objective truths. Every book is a product of culture, but many books express objective truths. Every statement expressed in human language is a product of culture, but many such statements express objective truths. So too, a moral code could be a product of culture and yet still express some objective truths about how people ought to live.

Against premise 2, it's too simple and vague just to say "*All moral beliefs are products of culture*." Yes, culture greatly influences moral beliefs, even though we may reject some cultural norms. But other factors contribute much to our moral thinking, such as:

- *Individual differences*. Individuals, based on their own thoughts, feelings, and experiences, may form moral beliefs that diverge from those of the wider group.
- *Logic*. When society teaches us a pair of logically incompatible norms, rationality may push us to reconsider both norms and reject one.
- *Biology*. To promote our survival, evolution built into us social instincts (like concern for the group and the golden rule), as it did for other social animals (like wolves and bees).
- *Religion*. Genuine religion can call us to a higher standard than does society, for example, to love everyone instead of just our group.
- *Developmental psychology*. Lawrence Kohlberg claimed that people of every culture develop moral thinking through set stages. CR is an immature conformism typical of teenagers; as we grow out of this, we begin to think for ourselves more rationally about morality. So not all social scientists accept CR; many oppose it.

CR is one-sided in considering only culture and ignoring other factors (like individual differences, logic, biology, religion, and developmental psychology). We'll use these other factors as we later construct our alternative to CR.

No Neutral Standpoint

Here's our third argument—and a stricter formulation:

> (3) My *no-neutral standpoint* argument points out that there's no neutral rational standpoint for criticizing another culture's values. Experiments can decide scientific issues; if someone thinks that heavy objects fall faster than light ones, we can drop objects of different weights and see which hits the ground first. Moral issues aren't like this. When we argue about ethics, we just assume the values of our own culture. There's nothing objective here.
>
> No belief that can't be evaluated on culturally neutral grounds is objectively true.
> All moral beliefs are beliefs that can't be evaluated on culturally neutral grounds.
> ∴ No moral beliefs are objectively true.

Again, both premises are doubtful. Against premise 1, suppose that I believe that it rained on this exact spot exactly 500 years ago today. This belief could be objectively true, even though there are no culturally neutral ways to evaluate it. Only a small percentage of all truths are knowable and arguable. So there could be objective moral truths, even if we had no solid way to know them or argue about them.[10]

Against premise 2, there may be strong, transcultural ways to criticize ethical beliefs (ways that don't just criticize **their** cultural beliefs by appealing to **our** cultural beliefs). We'll start on this in the next section. And we'll learn that, yes, ethics can have a strongly rational dimension—one that CR totally misses.

Logic, Cultural Values, and Consistency

A big problem with CR is that a culture's values may have inconsistencies. A culture (like the United States) may hold that *all people* are created equal and have equal rights, while also denying important rights to *black people* (black slaves) and to *women people* (who couldn't vote before 1920). Or a culture may hold that *all people* ought to be treated with respect and consideration, while also teaching that it's all right to mistreat *Moslem people* or *gay people*.

It's unclear what cultural relativists should say about such cases. Can contradictory value judgments (like "Slavery is wrong" and "Slavery is not wrong") both be true, if each accords with some (mutually inconsistent) socially accepted values? By standard logic, views that lead to self-contradictions are thereby wrong (remember the indirect proofs of high-school geometry). So then, since CR leads to self-contradictions, thus CR is wrong.

Perhaps all cultures have moral inconsistencies; often higher cultural values conflict with lower ones. I see *higher values* as those closely based on the common core of objective values that most cultures tend to share (like the golden rule and the norms to be consistent and get our facts straight), and *lower values* as based on less rational forces (like conformity, ignorance, and hating those who differ from us). This struggle between higher and lower values can drive *moral progress* in a society.[11]

Inconsistencies, in factual beliefs or moral beliefs, are *bad* and *ought to be avoided as far as possible*. This is a rough statement of the **consistency norm**, which I take to be an objective value that most cultures tend to share. CR claims that we can criticize a culture's values only by assuming another culture's values, and so we cannot criticize cultural values in any objective way. But I claim that we can criticize a culture's values on the basis of consistency; a culture with inconsistent values is flawed in an objective way and ought to change at least one of the conflicting values.

---

[10] This is NOT going to be my view about ethics, but the *possibility* of this view raises problems for premise 1.

[11] On CR, the idea of *moral progress* makes no sense. Suppose that a society moves from A (racism) to B (equality for all); then A is better relative to earlier cultural norms, and B is better relative to later cultural norms—but neither is better in an unqualified (objective) sense.

Consistency's role in ethical thinking may appear to be small. But I'll argue that, since the golden rule is a consistency principle, consistency's role in moral thinking is enormous.

Logic, Cultural Values, and Reasoning[12]

Sometimes we can criticize a cultural value by showing that it rests on a false factual premise. Recall our Wonder Woman example:

> Diana, later to become Wonder Woman, was born on the island of Themyscira. This is the home of the Amazons, a tribe of women warriors. Themyscira is a monarchy ruled by a queen. In the Amazon culture, only women ought to rule.

The Amazons believed "*Only women ought to rule*" because they believed that having a man on the island would cause massive deaths among the Amazon women, who would then cease to be immortal. So their belief was based on an argument like this:

Having a man on the island would cause massive deaths.
If having a man on the island would cause massive deaths, then only women ought to rule on the island.
∴ Only women ought to rule on the island.

In the Wonder Woman story, a man (Steve Trevor, an American pilot) lands on the island—and this doesn't cause massive deaths. Assume that this shows that the first premise ("*Having a man on the island would cause massive deaths*") is false. And assume that the Amazonian belief that *only women ought to rule on the island* rests only on the above argument. Then we've shown that this Amazonian belief is defective, since it rests on a factually false belief.

False beliefs (like inconsistencies) are *bad* and *ought to be avoided as far as possible*. In forming our moral beliefs, we *ought* (and I take this to be an objective value shared by most cultures) to try to get our facts straight and base our moral thinking on genuine knowledge instead of falsehoods. CR claimed that we can criticize a culture's values only by assuming another culture's values, and so we cannot criticize the values of a culture in any objective way. But we can criticize a culture's values on the basis of factual errors (as well as inconsistencies); values based on factual errors are flawed.[13]

Recall our example about early America:

In early America (before roughly 1920), no women ought to vote or hold office.

---

[12] While this section criticizes *sexist* social norms, similar criticisms apply to *racist* social norms (as in American slavery and segregation, and German Nazism). See my *Ethics*, pp. 124–35, and *Ethics and the Golden Rule*, pp. 136–51.

[13] How the value is held (on the basis of factual errors) is flawed. In rare cases, there may be another argument, one with true premises and a valid form, that can establish the value; I don't deny this.

Did early America base "*No women ought to vote or hold office*" on further reasoning? Let's assume for the moment that it did. Many argued this way:

Having women vote or hold office would destroy the family.
If having women vote or hold office would destroy the family, then no women ought to vote or hold office.
∴ No women ought to vote or hold office.

The first premise is factually false (as became very clear after 1920, although it was clear to many people earlier). So the cultural belief that *no women ought to vote or hold office* is flawed if it's based just on this argument.

Here's another popular argument[14]:

No women are capable of sophisticated political thinking.
Only those capable of sophisticated political thinking ought to vote or hold office.
∴ No women ought to vote or hold office.

Premise 1 can easily be shown to be false.[15] Premise 2 would exclude men from voting if they're incapable of sophisticated political thinking. Premise 2 would lead to something like everyone, male and female, being given a test to measure their political sophistication—and then only those who do well would be permitted to vote, but this approach and premise 2 clash with strongly held American values.

I contend that any further reasoning for *no women ought to vote or hold office*, if carefully formulated, would have similar problems.

And so, in at least two ways so far, we can criticize cultural values on neutral terms—we can point out genuine defects in a value system instead of attacking it just because it differs from our own values:

- A pair of cultural values may conflict logically; such values violate the "be consistent" transcultural norm.
- A cultural value based on reasoning can be criticized if rests on a false factual premise; such values violate the "be informed" transcultural norm.

Let's now suppose that early America accepted this following norm as *basic* (as not based on further reasoning or on some particular view of the factual situation):

No women ought to vote or hold office.

Suppose, in other words, that early America had no further reasoning for this norm but accepted it as an *intrinsic cultural attitude*. Even then, we can test the

[14]The old cultural norm was universal that **no women** ought to vote or hold office. Logically, a universal conclusion requires a universal premise, like that **no women** are capable of sophisticated political thinking—which is absolutely silly.

[15]We could have equally motivated randomly chosen males and females in a political science course and see how well each group does. Both groups would do about the same (as males and females do about the same in my logic courses, despite the stereotype that males are much more logical than females).

culture's *consistency* in holding this norm. We can ask people in the culture questions like these:

- Look at our neighboring country in which a woman president does an excellent job. Suppose that the best person for the presidency in **your** country (who would do a much better job than the other choice) is a woman. Even then, you say, the woman ought not to be elected president? Why? Suppose that this woman candidate was **you**. In this case, you shouldn't be considered for the presidency (even though you're the best person for the job). Do you really believe this? Why?
- "No taxation without representation" pushed the American colonies to revolt from England. They complained that England taxed colonists who couldn't vote for members of Parliament. But the early United States violated this same idea, since it taxed women who couldn't vote for members of Congress. Can't you see the inconsistency here? It violates deep American values to tax women while denying them the vote.
- By American thinking (in the Declaration of Independence), all persons are created equal in rights, including the right to vote for their rulers. By this, how can it be right to deny the vote to women? Doesn't this violate Central American values?
- Suppose that you're a woman who is very intelligent, interested in politics, and qualified to vote wisely—while your husband is stupid, ignorant of politics, and votes foolishly. Is it right to allow your husband to vote while you are denied this right? Do you really think this? Why would this be so?
- We could also point to countries where women have an equal right as men to vote and be voted for—and how this works well and tends to work for the good of everyone, including women and families. Why do you reject this approach?

Denying women the right to vote and hold office is difficult to do consistently. The idea continued because of unthinking habit ("We've always done it this way"). So moral norms that are held in a basic way may still be argued about, on the basis of consistency.[16]

A Broader Consistency Norm

We ought to be consistent in thought and action. This, I claim, is an objective, transcultural norm that's useful in ethical thinking. *Consistent* here is taken broadly: it's about things like consistency in beliefs, ends and means, living in harmony with our moral beliefs, evaluating similar cases similarly, and treating others only as we're willing to be treated in the same situation (GR). These ideas are already widely accepted across cultures (in practice, if not in formulation); we need to clarify and

[16]When arguing ethical issues with those of a very different perspective, it's often useful, instead of arguing from *your* premises, to try to understand the other person's perspective and look for inconsistencies.

organize them, show how to use them in moral thinking, and defend them from various perspectives.[17]

First, *consistency in beliefs* demands that we not accept logically incompatible beliefs—and that we not accept a belief without also accepting its logical consequences. Suppose that Ima Relativist begins her essay by contradicting herself:

> Since morality is relative, *no duties bind universally*. What's right in one culture is wrong in another. Universal duties are a myth. So *everyone ought to be tolerant of others*.

Ima's first italicized statement ("No duties bind universally") is logically incompatible with her last ("Everyone ought to be tolerant of others"). Ima is confused and inconsistent. She violates the following consistency imperative[18]:

Don't combine these:

- I believe "No duties bind universally."
- I believe "Everyone ought to be tolerant of others."

Consistency requires that Ima give up one belief or the other, but it doesn't say which to give up. *Consistency norms forbid inconsistent combinations but don't tell us specifically what to believe or what to do.* So consistency norms, while giving us some rational guidance, keep us free to form our own beliefs.

Our criticism of cultural relativism appealed to consistency. Socrates taught us to evaluate a philosophical view by first understanding it clearly and then seeing whether we can consistently hold it. Very few people can consistently hold CR when they understand clearly how it applies to racism, global warming, teaching morality to children, tolerance toward other cultures, and independent moral thinking.

I've had a beard most of my life. When people say "All bearded persons are crazy," I like to challenge their consistency. I ask, "Did Lincoln have a beard?" They say *yes*. I ask, "Was he crazy?" They say *no*. I say, "You contradict yourself."

Don't combine these:

- I believe "All bearded persons are crazy."
- I believe "Lincoln is a bearded person."
- I believe "Lincoln isn't crazy."

---

[17] My "*We ought to be consistent in thought and action*" is compatible with a range of metaethical views. An "*inconsistency*" might be based on logic and the meaning of terms, or it might be a *disharmony* that violates a moral truth or is forbidden by God or socially disapproved. Details will follow. As in math and logic, the rough ideas here are mostly agreed upon, but the foundations are disputed. My treatment of moral consistency and the golden rule here is a shorter version of Chapter 7 of my *Ethics*. See also my *Formal Ethics*, *Ethics and the Golden Rule*, and *Introduction to Logic*; the last of these puts my consistency norms into a precise formal system, using much logical machinery.

[18] I prefer the brief imperative form "Don't combine …" I'd also assert longer forms, like "You ought not to combine …" or "It's bad to combine these inconsistent things: …"

A consistency challenge doesn't tell us exactly what to believe. It just says, "The things you accept don't fit together—so you must change something."

Abraham Lincoln often used consistency against slavery (Basler, 1953: 223–4). Suppose you think that having higher intelligence gives one a right to enslave another. Do you then think that anyone with higher intelligence than you has a right to enslave you? Very few would believe that.

Don't combine these:

- I believe "Anyone with higher intelligence has a right to enslave anyone with lower intelligence."
- I don't believe "If X has higher intelligence than me, then X has a right to enslave me."

This forbids accepting a principle while not accepting its logical consequences.

Consistency encourages a reflective equilibrium between our principles and our concrete judgments. If we accept a principle but reject a concrete judgment that follows from it, we have to change either our principle or our concrete judgment; so we're pushed to think about things more deeply. Reflective equilibrium is based on consistency.

Consistency norms forbid inconsistent combinations. They don't say to shun emotions, never change our minds, or prove all our beliefs (which is impossible). Consistency norms can be defended in various ways—perhaps as self-evident truths, divine commands, or social conventions. They're part of normative ethics (saying that we *ought* to be consistent) and also metaethics (giving conditions needed to be *rational* in our moral beliefs).

Consistency norms aren't exceptionless. They don't apply if we psychologically can't be consistent, if the logical connections are too complex for us to grasp, or if some stronger duty interferes (perhaps Dr. Evil will destroy the world unless we're inconsistent). All our consistency norms are subject to implicit qualifications (Gensler, 1996: 15–39).

Consistency doesn't guarantee truth (since it's possible to be consistent but wrong), but it often points us toward the truth. Inconsistencies can lead us to see that a person is lying. And finding inconsistencies in a view can help us discover a better view.

There's also *inconsistency in will*. Here's an example:

Don't combine these:

- I resolve to eat nothing.
- I eat this granola bar.

If I combine both parts, then I'm inconsistent and need to change something; consistency alone doesn't tell me specifically what to do, since that depends on the situation. If my medical procedure requires fasting, then I should resolve and not eat. But if my fasting is an unhealthy way to diet, then I should eat and not resolve.

Here are further examples of inconsistency in will. I make a firm resolution (to run every day), but then act against it (I put it off and don't do it). I have all-things-considered desires that I know are incompatible (e.g., to become a doctor and to party all the time). As a politician, I endorse incompatible proposals. Or I violate ends-means consistency or conscientiousness.

Ends-means consistency requires that we keep our means in harmony with our ends. We violate this if we (1) have an end, (2) believe that to fulfill this, we need to carry out certain means, and (3) don't carry out the means. Suppose Maria has the all-things-considered goal to become a doctor. She realizes that, to do this, she needs to study hard and get good grades, but she doesn't act accordingly. Ends-means consistency forbids a combination:

Don't combine these:

- I have the goal to become a doctor.
- I believe "Achieving this requires that I study hard and get good grades."
- I don't study hard and get good grades.

Since Maria's goals, beliefs, and actions don't fit together, she must change something. Maybe her doctor goal is unrealistic and should be rejected, or maybe she just needs to carry out the means. Consistency doesn't say what to change.

Ends-means inconsistency, like belief inconsistency, is common. We often do what's easy or immediately satisfying, instead of what's needed to fulfill deeper goals. While Aristotle defined "humans" as "rational animals," we're imperfectly rational, and our rational and animal dimensions can fight each other.

Some goals are evil. Our ends-means norms don't command evil means to promote an evil goal; instead, they just forbid inconsistent combinations, like:

Don't combine these:

- I have the goal to take maximal revenge against Al.
- I believe "Achieving this requires slashing Al's tires."
- I don't slash Al's tires.

You shouldn't combine these, because (1) the three are inconsistent, and (2) you shouldn't have this goal (and so you shouldn't combine this goal with other things). Consistency, again, forbids a combination; it doesn't say that, if you in fact happen to have this goal and this belief, then you ought to do the act. In addition, ends and means need to satisfy other consistency principles. Here the ends and means likely violate the golden rule (given that I'm not willing that my tires be slashed in similar circumstances) and clash with my other goals (e.g., to avoid a mutually destructive revenge-war).

In the movie *Erin Brockovich*, a power company in order to maximize profits (goal) released toxic water into the ground (means) without cleaning it up; they knew the toxic water would cause cancer to many—and they weren't willing that this be done to them in the same situation. After a law clerk investigated, the company had to pay a fine and clean things up. Their unrestricted goal (to maximize profits, even by means that harm others) and the means (releasing toxic water) fail the golden rule test. To be fully consistent, our goals and actions need to satisfy the golden rule; a morally responsible company would do this.

Immanuel Kant's formula (Kant, 1964: 96)—"Treat humanity, never simply as a means, but always at the same time as an end"—points to a difference between how we should treat things and how we should treat people. We can use a hammer as a *mere means* to promote our goals, but it's wrong to use others as a *mere means*. It can be fine to use a person as a means, as I use a dentist to get healthy teeth. But, at the same time, I must treat persons as persons (and not as things) and take into account what happens to them, as I want others to do toward me; I must treat others only as I'm willing to have myself treated in the same situation. It's evil to treat persons as a *mere means* to promote my goals.

Ends and means are important to practical reason and human life. We have many goals—including food, shelter, health, companionship, and meaningfulness. Practical reason has us try to understand our goals, investigate how to satisfy them, imagine various end results, satisfy ends-means consistency, and reject ends or means that lead us to violate the golden rule consistency.

Conscientiousness requires that we keep our actions, resolutions, and desires in harmony with our moral beliefs. We violate this if our moral beliefs clash with how we live and want others to live. Suppose I accept pacifism: "One ought never to kill a human being for any reason." If I'm conscientious, then (1) I never intentionally kill a human being, (2) I resolve not to kill for any reason (even to protect myself or my family), and (3) I don't want others to kill for any reason. Similar requirements cover beliefs about what is "all right" ("permissible"). If I'm conscientious, then I won't believe that something is all right without consenting to (approving of) the idea of it being done, and I won't do something without believing that it would be all right for me to do it.

Conscientiousness says "Avoid inconsistencies between your moral beliefs and how you live and want others to live." Here's an example:

Don't combine these:

- I believe "I ought to do A now."
- I don't act to do A now.

When we combine these, our moral belief clashes with our actions—and consistency requires that we change one or the other. Consistency doesn't say "Our conscience is always right"; if our moral belief clashes with our actions, it could be that our actions are fine but our moral belief is wrong. Here's a consistency analogue of "Practice what you preach":

Don't combine these:

- I believe "Everyone ought to do A."
- I don't act to do A myself.

This doesn't assume that our norms are correct (so if we preach universal hatred then we ought to hate). Instead, it forbids inconsistencies between our norms and our actions; if these clash, then something is wrong (maybe our norms).

We can use consistency to criticize norms, even ones that seem self-evident. Suppose that my culture taught me to enjoy beating up short people and to accept *shortism*: *"All short people ought to be beat up, just because they're short."* To accept shortism consistently, I have to believe that if I were short then I ought to be beat up—and have to *desire* that if I were short then I be beat up. Consistency forbids this combination:

Don't combine these:

- I believe "All short people ought to be beat up, just because they're short."
- I don't desire that if I were short then I be beat up.

I'd likely violate this and be inconsistent (especially if I know what it's like to be beat up and imagine myself, vividly and accurately, in the place of short people treated this way). This same consistency approach can criticize other discriminatory principles (racial, religious, gender, sexual orientation, etc.).

My approach here leaves open what sort of "inconsistency" is involved here. Suppose that my moral beliefs clash with how I live. R.M. Hare's prescriptivism[19] says I violate logical consistency (misusing the term "ought"). But perhaps I instead violate a conscientiousness duty to keep my moral beliefs in harmony with how I live—a duty perhaps based on social conventions, divine commands, or self-evident truths. Then violating conscientiousness involves an objectionable clash between my moral beliefs and how I live, but not necessarily a logical inconsistency or self-contradiction.

Impartiality requires that we make similar evaluations about similar actions, regardless of the individuals involved. If we're impartial, we evaluate an action based on what the action is like—and not based on who plays what role in the situation. If we judge that an action is right (or wrong) for one person to do, then we judge that the same action would be right (or wrong) for anyone else to do in the same situation.

I violate impartiality if I make conflicting evaluations about actions I regard as exactly similar or relevantly similar. Two actions are *exactly similar* if they have all the same properties in common. They're *relevantly similar* if the reasons why one fits in a given evaluative category (good, bad, right, wrong, or whatever) also apply to the other. In the actual world, no two actions are ever exactly similar (have all the same properties in common). But the notion applies to hypothetical cases. To test my impartiality, I can imagine an exactly similar action where I'm on the receiving end of the action.

Here's my Good Samaritan example (Luke 10: 30–35). Suppose that, while jogging, I see a man who's been beaten, robbed, and left to die. Should I help him, perhaps by running back to make a phone call? I think of excuses why I shouldn't; I'm busy, don't want to get involved, and so on. I say "It would be all right for me not to help him." But then I consider an exactly reversed situation. I imagine myself in his place; I'm the one beaten, robbed, and left to die. And I imagine him in my place; he's jogging, sees me, and has the same excuses. I ask, "Would it be all right for this man not to help me in this situation? Surely not!" But then I'm inconsistent.

[19] See his *Freedom and Reason* (New York: Oxford, 1962). Hare had a big influence on my thinking.

What's all right for me to do to another has to be all right for the other to do to me in an imagined exactly reversed situation.

In the actual world, no two actions are ever exactly similar. But we can always imagine an exactly similar action where I'm on the receiving end of the action. I violate impartiality if I violate this principle:

Don't combine these:

- I believe "It would be all right for me to do such and such to X."
- I believe "In an exactly similar situation, it would be wrong for X to do this to me."

This sounds a little like the golden rule. But it's about impartiality, about making similar evaluations about similar actions. The genuine golden rule is about actions and desires ("Treat others as you want to be treated"), not about evaluations.

My example uses an imagined "exactly reversed situation" where all my properties are switched with those of the other person. Let me explain this further. Suppose we list my properties and those of the other person ("X"):

My properties: jogging, very busy, has blue eyes, …
X's properties: beaten and robbed, needs a doctor, has brown eyes, …

Imagine that the list contains all our properties, even complex ones; the list would be too long to write out—perhaps infinitely long. When I imagine an exactly reversed situation, I imagine the list of properties being reversed:

My properties: beaten and robbed, needs a doctor, has brown eyes, …
X's properties: jogging, very busy, has blue eyes, …

Here I'm beaten and robbed, and X is the jogger. We also reverse relationships; so if X helped me in the past, I'd imagine that I helped X in the past.

Instead of switching all the properties in my mind, I can switch just the ones relevant to evaluating the action. If I'm not sure if a property is relevant, I can switch it anyway, to be safe. Then I imagine a "relevantly similar" situation.

Suppose I'm driving and see a hitchhiker. Should I pick him up? If I don't, he may spend a long time waiting; I know what this is like from when I've hitchhiked to backpacking trailheads. On the other hand, people who pick up hitchhikers are sometimes robbed or hurt. Impartiality tells me that whatever judgment I make on my picking up the hitchhiker (that it's obligatory, wrong, or neutral), I must make the same judgment on the imagined reversed-situation action. Impartiality doesn't tell me what to do, and here it doesn't push me toward an obvious answer. Rather, it encourages me to reflect on the action from both perspectives (mine and the hitchhiker's). And it insists that, whatever I decide, I must apply the same standards to myself that I apply to others.

These examples test our impartiality by seeing how we evaluate an *imagined* second case. The next example uses an *actual* second case that's recognized to be relevantly similar. This example also shows how such reasoning can help us to recognize duties toward ourselves.

In the movie *Babe*, the athlete Babe Didrikson (1914–1956) needed a colostomy operation to save her life. From fear, she decided that she shouldn't have it. But her husband, thinking that she should have it, had her talk with another woman in similar circumstances, who also had to choose between dying and having the operation. Babe instinctively told the woman, "You ought to take courage and have the operation—for life is our greatest gift." But then Babe realized that she had to apply the same principles to herself that she applied to others. So she decided that she too ought to have the operation.

Babe at first violated this consistency imperative:

Don't combine these:

1. I believe "I ought not to have the operation."
2. I believe "You ought to have the operation."
3. I believe "Our cases are relevantly similar."

She held (1) because she feared the operation's results, she held (2) because she thought life is our greatest gift, and she held (3) because she thought that any reasons that justify one operation would also justify the other. Since her beliefs were inconsistent, she had to reject (1), (2), or (3). She in fact rejected (1), saying that she too ought to have the operation, just as the other woman in her similar situation should have it. She could have rejected (2), saying that neither should have the operation; she didn't do this, because she so strongly believed that the other woman ought to have it. Or she could have rejected (3), saying that the operation was right in one case but not the other, because of such and such differences, but she couldn't think of reasons that justify one operation but not the other. So consistency, while not telling her what exactly to believe, helped her to form her beliefs.

Some object that appealing to relevantly similar actions is slippery. What keeps you from appealing to trivial differences—from saying "It's all right for me to kill you but wrong for you to kill me, because I happen to have six toes and you don't"? If you pick trivial differences, we can appeal to hypothetical cases. Imagine a case where I have six toes (instead of you). Do you really think that in this case it would be all right for me to kill you? No one would believe this. In appealing to relevant differences, we have to give the factor equal weight regardless of which side is imagined to have it; this isn't easy to satisfy. At times, though, it's cleaner to appeal to imagined exactly similar cases.

Consistency norms respect our moral freedom, since they don't tell us specifically what to do or believe. They also promote moral rationality, since they guide us on how to work out our views consistently; they forbid, for example, criticizing others for doing certain things without also criticizing ourselves when we do the same things in similar circumstances.

I've spoken of impartiality as a type of "consistency" between our evaluations. I'm leaving it open whether to take this as "logical consistency," or more generically as "uniformity." Suppose I make conflicting evaluations about similar actions. R.M. Hare's prescriptivism says that I violate logical consistency (since I misuse the term "ought"). Other views may say that I violate an impartiality duty to make similar evaluations about similar actions; this duty might perhaps rest on social conventions, personal ideals, divine commands, or self-evident truths. On these views, violating impartiality involves an objectionable clash between moral evaluations but not necessarily a logical inconsistency or self-contradiction.

Impartiality, as I use the term, requires that we make similar evaluations about actions we regard as exactly or relevantly similar; it doesn't say these things:

- "If it's all right for my sister to drive, it's all right for me to drive." This needs a clause about the actions being relevantly or exactly similar; maybe your sister has a license but you're too young to get one.
- "Always act the same way in the same kind of situation." It's fine to eat carrots one day and celery another day in a similar situation, both actions being neutral (all right either to do or to omit doing). We violate impartiality only when we make conflicting evaluations about actions we regard as relevantly or exactly similar.
- "Treat everyone the same." We may give more help to one who needs it (as in the Good Samaritan case) or prescribe different treatments for patients with different illnesses. These needn't involve making conflicting evaluations about similar actions.
- "Love everyone equally." This would destroy friendships and families. Suppose we love our children more than we love strangers—and we think it would be all right for any parent in similar cases to do the same thing. Then we're making similar evaluations about similar actions, and thus we satisfy impartiality.

On the last point, some forms of utilitarianism require *strong impartiality* ("We ought to have equal concern for everyone's good"). I require only *weak impartiality* ("We ought to make similar evaluations about similar cases"); this lets me believe that *I* ought to have greater concern for *my* children so long as I believe that in similar cases *others* ought to have greater concern for *their* children.

The Golden Rule

The golden rule (GR) says "Treat others as you want to be treated." GR is a global standard, endorsed by nearly every religion and culture, important for families and professionals across the planet, and a key part of a global ethics movement.

Here's my favorite story to introduce GR.[20] There once was a Grandpa who lived with his family. As Grandpa grew older, he began to slobber and spill his food; so the family had him eat alone. When he dropped his bowl and broke it, they scolded him and got him a cheap wooden bowl. Grandpa was so unhappy. Now one day the young Grandson was working with wood. "What are you doing?" Mom and Dad

[20] Grimm, Brothers (1812), "The old man and his grandson," http://www.gutenberg.org/ebooks/2591. See also http://harryhiker.com/goldrule.htm (my GR page) and http://harryhiker.com/stories.htm (GR stories).

asked. “I’m making a wooden bowl,” he said, “for when you two get old and must eat alone.” Mom and Dad then looked sad and realized how they were mistreating Grandpa. So they decided to let him eat with the family and to keep quiet when he spills his food.

The heart of the golden rule is switching places. You step into another’s shoes. What you do to Grandpa, you imagine being done to you. You ask, “Am I willing that if I were in the same situation then I be treated that same way?”

GR seems simple. But loose GR wordings invite objections; many academics dismiss GR as a folksy proverb that self-destructs when analyzed. I think we just need to understand GR more clearly. I put my improved wording on a shirt.[21] It has “the golden rule” with symbols for eight GR religions (Bahá’í, Buddhism, Christianity, Confucianism, Hinduism, Islam, Judaism, and Taoism). It also has my GR formula, which is intended to help us apply GR to difficult cases.

Treat others only as you consent to being treated in the same situation.

My GR formula commands consistency. It demands harmony between my action toward another and my desire about how I’d be treated in the same situation. It doesn’t replace other moral norms or theories, or give all the answers. It doesn’t say specifically what to do (so it doesn’t command bad actions if we have flawed desires). Instead of directly specifying right and wrong, it helps us figure this out for ourselves—by forbidding inconsistent combinations:

Don’t combine these:

- I do A to another.
- I’m unwilling that if I were in the same situation then A be done to me.

GR, far from being vague, is a precise consistency test. Suppose I force Grandpa to eat alone. I switch places in my mind: I imagine that I’m forced to eat alone in the same situation. Do I condemn this same action done to me? Then I condemn how I

[21] See https://www.zazzle.com/store/harrygensler

treat Grandpa. I condemn how I treat another if I condemn the same action when I imagine it done to me in the same situation.

To apply GR, imagine yourself in the other person's place on the receiving end of the action. If you act in a given way toward another but are unwilling to be treated that way in the same situation, then you violate GR. To lead reliably to right action, GR consistency needs to build on *knowledge* (of how your action affects another) and *imagination* (visualizing yourself in another's place).

GR follows from conscientiousness and impartiality. Suppose we want to steal Detra's bicycle. And suppose we're conscientious (keep our actions and desires in harmony with our moral beliefs) and impartial (make similar evaluations about similar actions). Then we won't steal her bicycle unless we're also willing that if we were in the same situation then our bicycle be stolen.

If we're conscientious and impartial, then:

We won't do A to another unless we believe that doing A would be all right.
We won't believe that doing A would be all right unless we believe that if we were in the same situation then it would be all right for A to be done to us.
We won't believe that if we were in the same situation then it would be all right for A to be done to us unless we're willing that if we were in the same situation then A be done to us.[22]
∴ We won't do A to another unless we're willing that if we were in the same situation then A be done to us.

So if we're conscientious and impartial, then we'll follow GR. Now we've been assuming that we *ought* to be conscientious and impartial. But then we *ought* to follow GR: we *ought* to treat others only as we consent to being treated in the same situation. So my GR isn't basic; instead, it's a theorem provable from the conscientiousness and impartiality requirements, which are part of my comprehensive "One ought to be consistent" norm.

Three Technical GR Distinctions

Three parts of my GR formula are especially important.

1. The first part is "in the same situation."

Don't combine these:

- I do A to another.

[22] On my approach, believing that something is *all right* (permissible) commits us to *consenting* to the act being done, or *being willing* that the act be done (where I use these as equivalents). But what exactly do these two terms mean here? I propose that these express a kind of legislating in our minds in which we, as a minimum, don't condemn, object to, disapprove of, forbid, protest, prohibit, or repudiate the action (and we thus permit, allow, agree to, condone, or tolerate the action—in some sense of these terms).

- I'm unwilling that if I were *in the same situation* then A be done to me.

People who reject GR usually understand it crudely, often as what I call the *literal GR* (*"If you want X to do A to you, then do A to X"*). By this, if you want Lucy to be kind to you, then be kind to her, and if you want Adam not to hurt you (or rob you, or be rude to you), then don't do these things to him. These seem sensible. The literal GR generally works well if you and X are in similar situations and if you have good desires about how you're to be treated. But if either condition fails, the literal GR can tell you to do crazy or evil things:

- *Different situations.* "If you want your doctor to remove your appendix, then remove your doctor's appendix."
- *Flawed desires.* "If you want others to hurt you [suppose you do], then hurt them."

We can avoid these problems by wording GR more carefully.

First, you and the other person may be in very different situations. Consider this instance of the literal GR:

> To a son who hears well but has a father with bad hearing: If you want your father not to speak loudly to you [since your hearing is normal], then don't speak loudly to him.

This ignores differences between you and your father; a same-situation clause fixes the problem. Ask this: "How do I desire that I'd be treated if I were in the same situation as my father (and thus hard of hearing)?" You desire that if you were in his same situation, then people would speak loudly to you; so you speak loudly to him.

We can take the "same" situation here as "exactly similar" or "relevantly similar." In the first case, I imagine myself in my father's exact place (with all his properties). In the second, I imagine myself having those properties of my father (such as being hard of hearing) that I think are or may be relevant to deciding how to speak to him; if you're unsure whether a property is relevant, then switch it anyway, just to be safe. Both approaches work fine.

Here's another case where the literal GR leads to problems:

> To a patient: If you want your doctor to remove your appendix, then remove your doctor's appendix.

Again, we need a same-situation qualifier. The patient clearly doesn't desire that if he were in his doctor's place (with a healthy appendix), then his appendix be removed by a sick patient ignorant of medicine. As you apply GR, ask this: "Am I willing that if I were in the same situation then this be done to me?" The other person's situation includes likes and dislikes. So if you're a waiter who hates broccoli, but your customer likes and orders it, then you imagine being served broccoli in a hypothetical situation where you like and order it.

We need a "same-situation" qualifier because people may have different needs and circumstances (and because GR is based on impartiality, which needs this same qualifier). The literal GR ignores such differences and has us treat everyone the same way, in accord with *our* needs and circumstances.

The *literal GR fallacy* assumes that everyone has the same likes, dislikes, and needs that we have. Here's a story to illustrate the fallacy. There once lived a monkey and a fish. The monkey followed GR, always trying to treat others as he wanted to be treated. But he sometimes applied GR foolishly. Now one day a big flood came. As the threatening waters rose, the monkey climbed a tree to safety. He looked down and saw a fish struggling in the water. He thought, "I wanted to be lifted from the water." And so he reached down and grabbed the fish from the water, lifting him to safety on a high branch. Of course, that didn't work. The fish died.

The monkey applied GR literally: treat others as you want to be treated. He wanted to be lifted from the water, so he lifted the fish from the water. He didn't consider how monkeys and fish differ. Being lifted from the water saves a monkey but kills a fish. So the monkey applied GR foolishly.

There was also a wise GR monkey, named Kita, who considered taking a fish out of the water. But Kita knew that this would kill the fish. As she imagined herself in his situation, she asked, "Am I now willing that if I were in the same situation as the fish, then I be taken from the water?" She answered, "Gosh no: this would kill me!" So she left the fish in the water. Since people (and animals) may differ, this same-situation clause is crucial; we are to treat others only as we consent to being treated *in the same situation.*

"KITA" is also an acronym (Know-Imagine-Test-Act) for some main elements for using GR wisely:

- Know: "How would my action affect others?"
- Imagine: "What would it be like to have this done to me in the same situation?"
- Test for consistency: "Am I willing that if I were in the same situation then this be done to me?"
- Act toward others only as you're willing to be treated in the same situation.

To lead reliably to right action, GR consistency needs to build on things like knowledge, imagination, creativity, rationalized desires, and a healthy self-love.

My favorite historical GR example is a civil-rights speech by President John Kennedy (June 11, 1963),[23] during the first black enrollment at the University of Alabama. While Kennedy didn't know about GR monkeys, his speech followed KITA:

- He first got people to *know* how blacks were being treated, in areas like education, employment, and voting.
- He had whites *imagine* themselves being treated that way on the basis of their skin color.
- To *test* their consistency, he asked whether they'd be content with being treated that way.
- He urged *acting* on GR: "The heart of the question is whether all Americans are to be afforded equal rights and equal opportunities, whether we are going to treat our fellow Americans as we want to be treated." He called for changes in actions, attitudes, and laws (including the 1964 Civil Rights Act).

---

[23] Https://www.americanrhetoric.com/speeches/jfkcivilrights.htm

The heart of morality is GR. And the heart of GR is switching places. What we do to Grandpa (or blacks, gays, or whomever we mistreat) we imagine being done to ourselves. And to avoid the literal GR fallacy, we can imagine ourselves in the other's exact place (having their likes, dislikes, needs, and so on).

2. The "unwilling that if" part of my GR formula is also very important:

Don't combine these:

- I do A to another.
- I'm *unwilling that if* I were in the same situation then A be done to me.

GR is about our present reaction to a hypothetical situation; it isn't about how we'd react if we were in that situation.

Suppose you're a nurse about to give a shot to a baby. The baby doesn't want the shot but needs it to neutralize a bee sting that would otherwise bring great pain or death. To apply GR, you imagine yourself in the baby's place. You imagine this situation: You're a baby who doesn't want the shot but needs it to neutralize a bee sting that would otherwise cause great pain or death. What do you now want to happen in this situation? Clearly, you'd now say "Give me the shot." So you're now willing that if you were in this situation then you'd be given the shot (even against your will). GR has you ask this question: "Am I now *willing that if* I were in the same situation as this baby then I be given a shot?" The answer is *yes*. So GR, correctly understood, lets you give the baby the shot.

People often ask the GR question wrongly, which forces them to do whatever the other person wants. They ask, "If I were in the baby's place, how would I then want to be treated?" Now if you were in the baby's place, then you wouldn't want to be given the shot; misapplying GR, we'd conclude that you shouldn't give the baby the shot. So it's better to apply GR as explained above. I can give the baby the shot (to protect it from great pain or death), since I'm now willing that if I were in her situation then I be given the shot. In asking the GR question, it's important to say "willing that if":

> Am I *willing that if* I were in the same situation then this be done to me?

Immanuel Kant's objection to GR[24] (Kant, 1964: 97) rests on this confusion. Here you're a judge, about to sentence a dangerous criminal to jail. The criminal protests and appeals (incorrectly) to GR: "If you were in my place, you wouldn't want to be sent to jail; so by the golden rule you can't send me to jail." You should respond: "I can send you to jail, because I'm now *willing that if* I were in your place (as a dangerous criminal) then I be sent to jail." You could add, "If I do such things, then please send me to jail too!"

---

[24] My *Ethics and the Golden Rule* answers 33 objections to GR. For example, 30 objects that GR appeals to selfish self-interest while 31 objects that GR appeals to an unrealistic pure altruism; both are wrong—GR says nothing about motivation and is compatible with either self-interest or altruistic motivation.

Sometimes we need to act against what others want. We may need to give a shot to babies who don't want it, refuse salespersons who want to sell us overpriced products, fail students who don't work, defend ourselves from attackers, or jail dangerous criminals. GR lets us act against what others want, if we're now willing that if we were in their situation then we be treated similarly.

3. The "don't combine these" part of my GR formula is also very important.

Don't combine these:

- I do A to another.
- I'm unwilling that if I were in the same situation then A be done to me.

GR forbids an inconsistent combination; it doesn't say which individual actions are right or wrong.

Recall that the literal GR can lead to absurdities in two main ways: We dealt with the different-circumstances problem by adding a same-situation clause. A second problem is that the literal GR can tell us to do bad things if we have flawed desires about how we're to be treated. I'll give four examples.

(A) There once was a woman named Electra. Electra wanted to follow GR, but she got her facts wrong; she thought electrical shocks were pleasant. Since she wanted others to shock her, she applied the literal GR and shocked them: "If you want others to give you electrical shocks, then give them electrical shocks." Given flawed desires, the literal GR can command evil actions.

We'll use a triple defense against flawed desires.

(i) My improved GR formula, instead of telling us specifically what to do, just forbids a combination:

Don't combine these:

- I give electrical shocks to another.
- I'm unwilling that if I were in the same situation then electrical shocks be given to me.

Since my improved GR doesn't say specifically what to do, it doesn't tell Electra to do evil things (like shock others).

(ii) GR consistency, to lead reliably to right action, needs to build on other things, like knowledge and imagination. If we're misinformed, then we might do evil things while satisfying GR consistency. Here Electra shocks others (an evil thing) but satisfies GR consistency (she's willing that she be shocked in similar cases), since she's misinformed and thinks these shocks are pleasurable.

(iii) We need to use reason against flawed desires. Here we'd show Electra that electrical shocks are painful (perhaps by giving her a small one). Once she understands this, GR consistency will lead her away from shocking others.

(B) Or suppose Mona hates herself and wants others to hate her; then the literal GR tells her to hate others. (i) But again, the correctly formulated GR forbids a combination but doesn't tell her to hate others. (ii) GR consistency, to lead reliably to right action, needs to build on other things (like knowledge, imagination, and here a healthy self-love). (iii) We can use reason against Mona's flawed desires. We can try to help Mona understand why she hates herself and how to neutralize this hatred—by not fixating on her negatives, by seeing herself and her good points in a more balanced way, and, if she's a believer, by appreciating how God loves her. Once Mona regains a healthy self-love, GR consistency will lead her more readily to love others.

(C) Or suppose Mike is a masochist who gets athletic satisfaction from pain and wants others to cause him pain; then the literal GR tells him to cause pain to others. My improved GR would handle this much like how it handles the Mona case. But here the same-situation clause is also important, since Mike is likely unwilling that he be caused pain if he were in the place of his victims (who presumably get no satisfaction from pain).

(D) Or suppose Adolf is a Nazi who so hates Jews that he desires that he be killed if he were found to be Jewish; then the literal GR tells him to kill others if they're found to be Jewish. Again, we can make three points. (i) The correctly formulated GR forbids a combination but doesn't tell Adolph to kill Jews. (ii) GR consistency, to lead reliably to right action, needs to build on other things (like knowledge, imagination, and here rational desires). (iii) We can use reason against Adolph's flawed desires. When we try to understand why he hates Jews so much, we'll likely find that his hatred has its source in things that can be rationally criticized. Maybe Adolf thinks Aryans are superior to Jews and racially pure; we can criticize this on factual grounds. Or maybe Adolf was taught to hate Jews by his family and friends, who hated Jews, called them names, and spread false stereotypes about them. Then his anti-Jewish desires came from false beliefs and social conditioning; his flawed desires would diminish if he understood their origin and broadened his experience and knowledge of Jews in an open and personal way. With greater rationality, Adolf wouldn't desire that he'd be killed if found out to be Jewish—and GR would be a powerful tool against his racism.

While this example was about a Nazi, the same idea applies to those who desire that they be mistreated if they were black, female, gay, or whatever. Such desires are likely flawed (as based on a social conditioning that uses false beliefs and stereotypes) and would be given up if we expanded our knowledge and experience of the group in an open and personal way.

The related *easy GR fallacy* assumes that GR gives an infallible test of right and wrong that takes only seconds to apply. Here's a story. Imagine a rich coal mine owner named Rich, who pays his workers only $1 a day. He's asked if he's willing to be paid only $1 a day in their place. He replies, "Yes, you can live well on $1 a day, so I'm willing that I be paid that in the place of my workers; I love GR—you just whip out this moral compass and in a few seconds you know whether you're acting rightly!"

Rich moves too fast. To lead reliably to right action, GR needs to build on knowledge and imagination, which may take time. Rich is willing that he be paid $1 a day in his workers' place (and so is consistent), but he's so willing because he thinks (wrongly) that his workers can live well on this much. If he knew how little $1 buys, he wouldn't think this. Rich needs to get his facts right. He should try going to the store to buy food for his family with only $1 in his pocket!

Now suppose that Rich decides to apply GR more adequately to how to run his mine. What would he do? Following KITA, he'd do four things.

- Rich would gain *knowledge*. He'd ask, "How are my company policies affecting others—workers, neighbors, customers, and so on?" To know this, Rich would need to spend time talking with workers and others.
- Rich would apply *imagination*. He'd ask, "What would it be like to be in the place of those affected by these policies?" He'd imagine himself as a worker (laboring under bad conditions for a poor salary), or a neighbor (with black smoke coming into his house). Or he'd imagine his children being brought up under the same conditions as the workers' children.
- Rich would *test* his consistency by asking: "Am I now willing that if I were in the same situation (as my workers, neighbors, or customers) then I be treated that same way?" If the answer is no, then his actions clash with his desires about how he'd be treated in a similar situation—and he must change something. To change company policies, Rich may need creativity to find alternatives. He might listen to ideas from others. He might learn what other companies and cultures do. He might imagine what policies make sense from a worker's perspective, explain current policies to a child, write an essay listing options, take a long walk, or pray about it. While GR doesn't tell him which policies to consider, GR gives a fairness test for any proposed policy; Rich must be able to approve of it regardless of where he imagines himself in the situation: as owner, worker, neighbor, or customer. The final solution, while maybe not satisfying everyone fully, needs to be at least minimally acceptable from everyone's situation.
- Rich would *act* on GR: "Treat others only as you consent to being treated in the same situation." Yes, it's a simple formula. But applying it wisely may require preparatory work on knowledge, imagination, and creativity. We'll never be perfect in these areas; but the fact that we may never do something perfectly doesn't excuse us from trying to doing it as well as we reasonably can.

The related *too-simple-or-too-complex GR fallacy* assumes that GR is either so simple that our kindergarten GR is enough for adult decisions or so complex that only a philosopher can understand it. On the contrary, GR is *scalable*. You can teach GR to small children ("Don't hit your little sister—you don't want us to hit you, do you?"), while adults can use it in complex decisions (like how to run a coal mine in a way that respects everyone's rights and interests). Our understanding of GR needs to grow as we mature.

People sometimes describe a situation to me and ask, "What would GR tell us to do in this situation?" This shows a misunderstanding of how GR works. Properly understood, GR doesn't directly tell us what to do, and so it doesn't give us the solution to our problem. We've got to propose our own solution (after studying the facts and imagining ourselves in the place of the various parties). Then GR can test our proposed solution for consistency (and fairness); any proposed solution must be one that we're willing to have followed regardless of where we imagine ourselves in the situation.

As a consistency norm (forbidding a combination but not directly telling us what individual action to do), GR isn't a direct criterion of right and wrong; and so GR isn't a rival to moral norms like "One ought not to steal." GR works at a different level. GR doesn't impose a rule on us from the outside but rather takes our own rule (e.g., "Don't steal from me!") and pushes us to apply it consistently to how to treat others. GR is much like "Don't contradict yourself." GR's role isn't to replace other ethical theories but to supplement them—by giving a consistency tool that's often useful. Most ethical theories recognize the role of consistency and so should be able to accept GR on this basis.

GR is golden (valuable) because it captures so much of the spirit behind morality. It counters self-centeredness and helps us to see the point of moral rules. It's psychologically sound and personally motivating, engaging our own reasoning instead of imposing answers from the outside. It promotes cooperation and mutual understanding. It criticizes culturally taught racist or sexist moral intuitions. It concretely applies ideals like fairness and concern. And it's a global wisdom, common to most religions and cultures. So GR makes a good one-sentence summary of morality.[25]

Practical Reason and Biology[26]

Many cultures tend to share these three objective *Norms of Practical Reason*:

- **Consistency**: We ought to be consistent in thought and action. This covers things like consistency in beliefs, ends and means, living in harmony with our moral beliefs, evaluating similar cases similarly, and treating others only as we're willing to be treated in the same situation (GR).
- **Knowledge**: False beliefs are bad and ought to be avoided as far as possible. As far as practically possible, especially in our deliberations and moral thinking, we need to know the situation, alternative moral views, and ourselves.
- **Imagination**: We need, especially in our deliberations and moral thinking, a vivid and accurate awareness of what it would be like to be in the place of those affected by our actions.

---

[25] GR applies nicely to practical areas like racism, sexism, and other forms of discrimination; global warming; moral education; how to treat animals; immigration; medical ethics; and business ethics. See *Ethics*, pp. 124–9, and *Ethics and the Golden Rule*, pp. 108–62.

[26] For more, see my *Ethics*, pp. 161–70, and *Ethics and the Golden Rule*, pp. 125–35.

More simply: "Be consistent (including GR-consistent), informed, and imaginative."[27] These three are the objective core of **practical reason** and give strong tools for reasoning about moral issues. I propose that a belief about values is *reasonable* (*wise*) if and only if it isn't defective in terms of practical reason.[28]

My version of **cultural objectivism** holds that cultures *tend to share a common core of objective values* (namely these three norms of practical reason). These values may be implicit in how a culture works, instead of having a clear formulation. And cultures, while recognizing these values, may at times violate them (as when an evil ruler spreads massive falsehoods). I'm not denying either of these. Instead, I'm claiming that the great majority of human societies would recognize these values, at least implicitly (in how they think about ethical issues). These norms would get high acceptance across the world.

So most humans have some motivation to reconsider their action if they believe any of these:

- "My action is based on factual errors."
- "I'd act differently if I had more information."
- "I'd act differently if I better visualized my action's effects."
- "I'm inconsistent."
- "I'd complain if someone did that to me."

Let's assume that this is so. But **why** is this so? What explains why these elements of practical reason are so widespread across human cultures? I'd claim that biology and evolution give the best explanation. Humans are a social species with unusual reasoning powers; since the norms of practical reasoning promote species and group survival, we evolved to have these norms built into us. In explaining how this works, I'll mostly follow the explanation of Charles Darwin (1809–1882) about how morality arose in humans.

Darwin's *Origin of Species* (Darwin, 1859) claims that current biological species evolved from earlier ones through *mutation* (a species randomly produces organisms with slight differences in inheritable features) and *selection* (features that promote survival tend to produce more offspring with these same features). Repeating the mutation-selection process millions of times produces new life forms. Evolution

[27] We could, if we liked, build "vivid awareness of the relevant facts" (understood in such a way that it includes my imagination condition) into an expanded knowledge condition. Then we'd have only two parts: *As rational beings, we ought, as far as practically possible in our decision making, to be consistent and to be vividly aware of the relevant facts.*

[28] Human reasonableness (wisdom) is more or less, since we never have a complete understanding of the situation. Reasonableness isn't the same as truth; on controversial issues, sometimes reasonable people can differ. Practical reason narrows the range of acceptable views (so not just anything goes), but it needn't bring complete agreement. I often add a fourth "other things" condition of further factors needed for reasonable moral beliefs (e.g., we need to be free to think for ourselves, instead of just conforming), but these "other things" are ultimately derivative feelings and character traits that support consistency, knowledge, and imagination—and so I skip them here. See my *Ethics*, p. 126, and *Formal Ethics*, pp. 151–52.

explains many facts about organisms: comparative structure, embryology, geographical distribution, and fossil records.

Darwin's later *Descent of Man* (Darwin, 1871) focuses on human evolution. Darwin claims that the moral sense (conscience), which is rightfully supreme over human action, is the most important difference between humans and other animals. Analogues of morality exist in other social animals, such as dogs, ants, and bees. Wolves work together in packs and help each other. Dogs sympathize with another's distress or danger and can show fidelity and obedience. Baboons follow a leader and enjoy being with each other. Many social animals have concern for the welfare of their offspring or comrades, even risking their own safety. With social animals, evolution encourages instincts and behaviors that benefit the group's survival.

Humans are social animals, living in families or groups. Like other such animals, we evolved social instincts. We instinctively enjoy helping others and are distressed by another's misery. We often show concern for others, especially offspring and members of our group. We value the approval of others, internalize group values, and follow the leader. Opposing our social instincts are anti-social impulses, like lust, greed, and vengeance.

Primitive human morality was mostly instinctive and had arbitrary taboos, little discipline, and little concern for those outside one's clan or tribe. But humans gradually developed their intellectual powers through observation, language, and reasoning. Morality became more rational, disciplined, and concerned with consequences. As human groups expanded, so did our circle of concern. As religion purified our social feelings and moral instincts, we struggled toward a higher morality, supported by reason and directed to the good of everyone, even the weak and animals. Our noblest attribute became a love for all living creatures. GR sums up this higher morality: "Treat others as you want to be treated."

Darwin's natural-history explanation of why we think morally emphasizes two gifts from evolution that promote our survival: *social instincts* and *reason*. Any social animal with strong social instincts would develop a moral sense as its reason becomes more developed. And so humans over many centuries developed from a narrow, instinctive concern for offspring and fellow tribal members toward a higher, rational GR morality of concern for everyone (as is conceptualized in my three norms of practical reason).[29]

Other studies suggest the wide presence of my three norms, for example[30]:

- Bloom's experiments about morality in infants, starting at about 3 months. Little babies definitely don't like puppets who hit other puppets.
- Kohlberg's study of people's stages of moral thinking as they develop in different cultures. The highest stage is about golden rule consistency.

---

[29] For a fuller treatment, see my *Ethics*, pp. 161–68, and *Ethics and the Golden Rule*, pp. 131–35.

[30] See Bloom (2013), Kohlberg (1972: 13–6, 1981, 1984), Festinger (1957), Pfaff (2007), and the Parliament of the World's Religions (1993) *Declaration Toward a Global Ethic*, http://www.weltethos.org/1-pdf/10-stiftung/declaration/declaration_english.pdf

- Festinger's study of *cognitive dissonance* (a stressful reaction to inconsistencies). Evolution programmed our minds to avoid inconsistencies, as it programmed our bodies to spit out poisons.
- Pfaff's study of ethics and neuroscience. Pfaff describes brain processes that go on when we apply the golden rule; he thinks that these started so that mothers would care for their babies, which is essential for humanity's survival.
- The global ethics document of the Parliament of the World's Religions (expressing a strong consensus of a great number of religions across the world) endorsed the golden rule as "the irrevocable, unconditional norm for all areas of life."[31]

Anthropology has grown to accept the existence of *ethical universals*, which are present in the great majority of cultures; the golden rule is often a central example. A summary of research about ethical universals says this:[32]

> Perhaps the most agreed-upon universal moral value is the Golden Rule. Several writers … have identified the many (only slight) variations of "Do unto others as you would have them do unto you" found within all of the major religions. The case for the universality of the Golden Rule is most impressive.

This research summary includes other universal values too (for example, to seek truth and justice, to act in accord with one's conscience, not to hurt others, and to be caring and compassionate and forgiving), but these can be justified through my framework.

The global ethics document (of the Parliament of the World's Religions, which I just mentioned as endorsing GR's centrality) endorses other values closely related to GR. It also endorses four general duties widely accepted across the globe in world religions—against stealing, lying, adultery, and killing—which protect four key human goods, namely possessions, speech, family, and life; these can be justified by GR and my framework.[33]

If we think of GR and the rest of practical reason as thus "written on the human heart" (Romans 2: 14–15), we shouldn't think of these as always written in big, bold letters. Sometimes the message is obscured by other things written on the human

---

[31] Also (see http://harryhiker.com/chronology.htm), the golden rule and its close relatives are present across the world in an impressively large number of sources (and often given as a summary of morality).

[32] See Kinnier et al. (2000: 4–16). The GR quote is from p. 5.

[33] My framework, instead of just listing transcultural ethical universals, tries to give the basic, transcultural norms that we can use to rationally evaluate and criticize values. I don't try to give a mechanism that will generate for all cultures the same answers to all concrete ethical questions (an impossible task). Consider the general imperative to protect life (which GR will endorse for humans, who very much want their own lives to be protected). Different cultures create different civil laws and moral rules to promote life—in areas like traffic laws and gun control, for example. Differences here are fine, if they accord with practical reason: consistent (include GR-consistent), informed, and imaginative; not all cultures need to work out the details the same way. But not just anything goes; we saw in sections "Logic, cultural values, and consistency", "Logic, cultural values, and reasoning", "A broader consistency norm", "The golden rule" and "Three technical GR distinctions" how to sharply criticize some cultural values.

heart—like selfishness, groupism, impulsiveness, stupidity, and cruelty. Morality can be a struggle. Despite this, I contend that the seeds of GR and the rest of practical reason are inside of us. We need to cultivate these seeds.

Metaethics and Objectivity

This chapter defends my version of "**cultural objectivism**," which holds that cultures tend to share a common core of objective values (namely our three norms of practical reason). But why should we follow these norms—and what kind of "objectivity" is claimed here? As mentioned earlier, cultural objectivism is compatible with a range of metaethical views. Let me give a few examples.

Intuitionism: The three norms of practical reason are *self-evident truths* (ethical truths that strike most of us as clearly true and continue to strike us this way as we investigate them further). They're "objective" in the sense of being *self-evident truths*. This is my view, but other views are possible.

My form of intuitionism accepts only self-evident norms about rationality. It's widely argued that moral norms can't be self-evident, because they're vague and widely disputed, but my three norms are clear and have a strong consensus behind them. The norm to avoid inconsistencies and falsehoods is presupposed in practically every area of thought, including science and math, and visualizing an action's effects is important in making choices. These three norms are more certain and more widely held than any premises that we might use to defend them. GR is widely held too, across different religions and cultures, and can be expressed in a clear way that resists objections; GR's correctness is clear in itself, even though it can be derived from consistency conditions that are more basic. An intuitionism based on these three norms of practical reason is the best form of intuitionism.

Cultural relativism: Except for perhaps very primitive societies, practically every society accepts "Be consistent (including GR-consistent), informed, and imaginative" and gives it a priority in moral and social thinking. These core values lead to social harmony, progress, and a better life for all, and they provide a needed rational structure for moral thinking and social planning. A *modified cultural relativism* might claim that what is "good" in a culture is *what the culture would approve of if it followed these norms of practical reasoning that are, because of our biological origins and rational progress, present and given priority in practically every non-primitive culture*. This approach, while defending moral reasoning, doesn't recognize transcendent moral truths; all morality is socially constructed or biologically inherited. The norms of practical reason are "objective" in the sense that they are present and given priority in practically every non-primitive culture, as grounded in our common human biology.[34]

[34] I'm not sure that this *modified cultural relativism* would work (I just invented it), and I certainly don't endorse it. I'm bringing it up to emphasize that the *cultural objectivism* that I'm defending in this chapter is compatible with a wide range of metaethical views (about the meaning and objec-

Emotivism and subjectivism: Morality is about your feelings; "X is good" means "Hurrah for X!" (emotivism) or "I like X" (subjectivism). Our three norms of moral reason can be based on your feelings in either of two ways. (1) If your feelings are *idealistic*, then you likely have positive feelings about being consistent (including GR-consistent), informed, and imaginative—and caring about others; then you accept these norms because they accord with your feelings. (2) If instead you're *egoistic*, then you can justify these three norms by appealing to self-interest:

- Inconsistencies lead to confusion and frustrated desires. To see this, imagine how miserable you'd be if whenever you believed or wanted something you also believed or wanted the opposite. You'd go crazy! Likewise, false beliefs and lack of imagination also lead to confusion and frustrated desires.
- Inconsistency brings a stressful condition that psychologists call *cognitive dissonance*. Evolution programmed our minds to avoid inconsistencies.
- Inconsistencies and false beliefs bring social penalties. They lead people to dismiss our ideas and cut us off from rational discussion. Society is especially harsh when we violate conscientiousness or impartiality; it trains us to feel guilt, anxiety, and the loss of self-respect over this.
- People mostly treat us as we treat them. So it generally pays in terms of self-interest to treat others well, as we want to be treated.
- GR promotes cooperation, which benefits everyone (including us) and brings social rewards. Selfishness promotes conflict, which hurts everyone (including us) and brings social penalties.
- GR ennobles us. It enhances our sense of *self-worth*, which is essential to happiness. We see ourselves as having worth because that's how we see everyone. If we violate GR, then we lose self-respect; if we treat others as worthless, how can we see ourselves, who are cut from the same cloth, as having worth?
- Students doing volunteer work typically report that they receive from others much more than they give to others. This gives evidence that helping others helps to make *us* happy.
- An experiment from Rimland (1982: 221–22) provides evidence that altruistic living promotes our happiness. Groups were asked to list persons they knew well and label them as happy or unhappy, and as altruistic or selfish. When responses are analyzed, "happy" people are almost always "altruistic" and almost never "selfish." So, judging by people's perceptions, self-interest supports GR.

So self-interest can justify our three principles fairly well too. The three principles of practical reason are "objective" in the sense that they are *based on widely shared feelings*; while our preference for ice cream flavors varies widely, our preference for true beliefs over false ones is widely shared.

R.M. Hare's prescriptivism: This works like emotivism except that it sees conscientiousness, impartiality, and GR consistency as built into the moral terms (and thus

---

tive status of ethical terms). To decide between these views is another task, but, again, I'd cast my vote for intuitionism.

a matter of logical consistency)—and it sees our three norms of practical reason as expressing prescriptions (or desires) instead of feelings.

Ideal-observer theory: "X is good" is a claim about what we'd desire under ideal conditions. If we use our three principles of practical reason to specify these ideal conditions, then "X is good" would mean "We'd desire X if we were fully consistent (including GR-consistent), informed, and imaginative." Then our three principles are justified by the definition of "good" and thus are "objective" in that they are *true by definition*.

Divine-command theory: "X is good" means "God desires X." Our three norms of practical reason can be justified as mirroring God's desires (and thus as "objective" in the sense of being *true*). The Christian Bible often condemns inconsistency; so Jesus (in Luke 13: 14–17) criticized hypocritical Pharisees whose actions clashed with their words. The Bible says much to support conscientiousness and impartiality. And practically every religion endorses GR, with many religions featuring GR as a summary of how to live.

So thinkers of diverse metaethical views could accept my approach to *cultural objectivism*—which holds that cultures tend to share a *common core of objective values*—and these are spelled out as *being consistent (including GR-consistent), informed, and imaginative*.

Religion, Ethics, and GR

We can study in two main ways how religion connects with morality (and GR). First, we can study religion as a part of culture, abstracting from its truth value; so far, this has been my approach. Second, we can study religion as participants and believers, as people (like me) for whom religion is a deep part of our lives and belief systems. This section does both.

First, as a part of culture, world religions across the ages have been important teachers of morality. While religions often differ about details (like dietary rules), they're more similar about the *objective core of ethics*, especially the golden rule (which all of the major religions and most of the minor ones support—see section "Practical reason and biology"). When religions differ about GR, they mostly give areas where different faiths can learn from each other. Here are examples:

- Hinduism emphasizes *Karma*, that the good or bad we do to others tends to come back to us; so treating *others* as we want to be treated tends at the same time to treat *ourselves* well. Karma is also present in other religions, like Christianity, but typically not with the same name and emphasis.
- Jainism emphasizes GR's wide scope, which applies to slavery and to our treatment of animals. The West was slow to learn these lessons.
- Buddhism talks about the misery we bring to ourselves when we act vengefully against enemies who treat us badly. Again, other faiths can learn from this.

- Islam has GR versions that stress *what to want for others* (we are to want for others what we want for ourselves) and *how we want God to treat us* (we are to treat others as we want God to treat us). Again, other faiths can learn.

For more, see my further writings.[35]

Religions mostly teach a *higher morality* than the general culture. As I noted in section "Logic, cultural values, and consistency":

> Perhaps all cultures have moral inconsistencies; often higher cultural values conflict with lower ones. I see *higher values* as those closely based on the common core of objective values that most cultures tend to share (like the golden rule and the norms to be consistent and get our facts straight), and the *lower values* as based on less rational forces (like conformity, ignorance, and hating those who differ from us). This struggle between higher and lower values can drive *moral progress* in a society.

On the other hand, religions can often be perverted from their original inspiration by twisted cultural standards. Jesus, Old Testament prophets, and many preachers today criticize those who follow religious externals but ignore the needs of their neighbors. Perverted religions can be anti-science, racist, and full of hatred toward other religions. Perverted religions can be very harmful, since religion can influence us strongly, for good or for evil.

While some have argued that the overall effect of religions on morality has been negative, I think that this view is one-sided and ignores the huge positive contributions of religion. I'd agree with the assessment of Bruce Sheiman, an atheist, about the positive social value of religion (Sheiman, 2009). His introduction claims: "By any empirical measure—defined in terms of theism's practical impact on individuals, society, and culture—religion is profoundly beneficial." Sheiman claims that strong empirical evidence shows that religious people, while accepting much the same moral norms as nonbelievers, tend to live out their morality better and be healthier and happier. Believers come out better by practically any measure: charitable giving, volunteering, donating blood, depression rates, suicide attempts, drug addiction, divorces, exercise, heart attacks, criminality, longevity, marital satisfaction, self-acceptance, moral purpose, peacefulness, personal happiness, and so on.[36] Sheiman sees religion as securing moral standards, teaching moral objectivity and universal love, and being the strongest force for morality and personal happiness. He claims that militant atheists who see religion as very harmful ignore massive empirical data to the contrary. And yet, he's an atheist; while seeing the tremendous value of religion, he cannot accept it himself.

For the rest of this section, I'll write as a believer (as a Christian, Catholic, and Jesuit). While I reject the *divine command theory* (discussed in the previous

---

[35] See my *Ethics*, pp. 154–61, and *Ethics and the Golden Rule*, pp. 34–67 and 76–107.

[36] See also Peter Glynn, *God: The Evidence* (Rocklin, Calif.: Prima, 1997, pp. 57–97), who mentions many further studies, and Lawrence Iannaccone ("Introduction to the economics of religion," *Journal of Economic Literature* 36 (1998), 1465–95), who disputes the idea that religion is gradually disappearing as people become more educated and scientific.

section), I do accept my own version of *natural law ethics* (a general approach that's popular with believers).[37]

In my approach to natural law, morality has three key dimensions: rational, biological, and spiritual. Atheists can accept the first two, but not the third. Believers and nonbelievers can share much about morality. Both can agree about practical reason—including the importance of being consistent (including GR-consistent), informed, and imaginative. Both can agree on how morality relates to our biological nature—including how we evolved concern for others and a sense of right and wrong; how humans ought to respect possessions, speech, family, and life; and how to use practical reason to explore ethical controversies. And both can be committed to morality and can cooperate to make the world a better place.[38] But atheists would reject morality's spiritual dimension, which understands morality in a religious context; the central idea here is that, *as spiritual beings called to eternal life with God, we ought, as far as practically possible, to love God with our entire being.*

The duty to love God with our entire being is comprehensive, covering all aspects of our lives. It requires that we obey God (whose will flows from His supreme wisdom and love); our other duties, given that God wants us to follow them, are also obedience duties. Thus obeying God fulfills all our duties. Religion personalizes morality. When we steal, lie, or break the golden rule, we do something that's wrong on nonreligious grounds, but we also disobey God and sin against the one who loves us and created us. (Sin is an intentional violation of God's will; atheists can accept wrongdoing but not sin.) So morality connects to our personal relationship with God.

Loving God with our entire being requires loving our neighbor and following the golden rule. Since God's commandments include "Love your neighbor" and "Treat others as you want to be treated," this is a corollary from the first point.

Loving God with our entire being requires that we praise and love God in every area of our lives. If you're an artist, praise God through your art. If you do science, praise God through your science. If you're a parent, praise God through your love for your family. If you hike the Grand Canyon, praise God through your hiking. If you struggle with faith, praise God through your struggles. Love, serve, and praise God with your entire being. For believers, morality is part of loving God. That's the biggest difference that theism makes to morality.

Religion gives a context for morality, a worldview that supports morality. We were created by a loving God to know right from wrong, to have free will, and to love one another. The core of right and wrong, the most important thing, is about caring for others, the golden rule, loving your neighbor, and how in living that way we can grow closer toward God. This provides a huge motivation for morality.[39]

---

[37] See my *Ethics*, pp. 165–70, and *Ethics and Religion*, Ch. 4–6.

[38] While believers and nonbelievers will mostly come to the same moral beliefs (for example, that stealing is wrong), there may be some differences. Believers will recognize a duty to worship God while nonbelievers won't. And there may be differences about issues like mercy killing, based on different beliefs about the origin and destiny of our lives.

[39] See the previous remarks in this section about the work of Bruce Sheiman.

From my perspective, nonbelievers miss much of the story about morality. Theistic ethics is more powerful because it appeals to the whole person. It sees GR, not only as a consistency imperative (rationality) and as a biological imperative for our individual and social well-being (biology) but also as a divine commandment and part of God's plan for us to grow in personhood toward our eternal fulfillment with him. And so we're motivated to follow GR by love of God, by gratitude toward God, and by seeing that we're all children of one God, made in his image and likeness. GR and morality fit religion nicely. It's possible to accept morality as an atheist, but then it's truncated—it omits the spiritual dimension, which can be an important part of moral thinking.

My Ethics Teaching in America and China

I've often taught introductory ethics, following the structure in my *Ethics* book. In the beginning, I give a strong and persuasive presentation of cultural relativism (like section "Pretending to defend cultural relativism") and then have my students write a one or two page reaction paper about the view; they are told not to do any further research but just to reflect on the view and write down their initial reactions—whether positive, negative, mixed, or confused. Most students find CR very interesting and initially very plausible; most like the view very much, but a few raise difficulties and objections. Then in class we demolish the view (like sections "Problems with cultural relativism", "Cultural differences", "Product of culture" and "No neutral standpoint").

The next part of this chapter diverges from what I do in my course.

- In this chapter (sections "Logic, cultural values, and consistency" and "Logic, cultural values, and reasoning"), I introduce two core objective values (about the role of consistency and knowledge in our moral thinking) by seeing how cultural relativism gets into trouble by neglecting them.
- In my course, I instead cover other metaethical views (subjectivism, ideal observer, divine command, intuitionism, emotivism, and prescriptivism); the reaction papers grow remarkably better over time, as students increasingly look for problems in a critical way instead of just following their initial impressions.

Then, in both this chapter and in my course, I cover what I call "the GR consistency view" (here sections "A broader consistency norm", "The golden rule", "Three technical gr distinctions", "Practical reason and biology", "Metaethics and objectivity" and "Religion, ethics, and GR"), which students like very much.

In summer 2019, I taught my ethics course in China (in English), at the University of Wuhan (yes, **that Wuhan**, but before the Covid pandemic hit). So far, I had only taught in America, and I knew well how US students react to cultural relativism (initially mostly positive, but very negative after they understand the objections) and the GR consistency view (very positive). But how would Chinese students react?

Would the views and issues even make sense to them? I didn't know, but I was dying to find out.[40]

On my first class day in China, I wore a Confucian golden rule shirt of my own design.[41] This is much like my other GR shirts (see section "The golden rule"), except that it has Confucius's GR saying (which he used to sum up his teaching about how to live) in Chinese and English. I told my students that we'd be studying Western ethics, yes, but also global ethics—and that our course's central GR idea is as rooted in their culture as it is in mine. My GR shirt had this graphic:

己所不欲勿施於人.
Confucius: What you do not want done to yourself, do not do to others.

I taught my course in China much like how I teach it in the United States, except that I connected it more with China:

- I studied Confucius and other classical Chinese thinkers (including some in the Daoist and Buddhist traditions), and often brought their ideas into the discussion.
- When we studied a controversy important in the West, I tried (often successfully) to find parallels in Chinese culture. For example, cultural relativism was represented in ancient Greece by Protagoras—and in ancient China by Teng Shih.
- I use Martin Luther King as an example of someone who went against cultural relativism, opposing racist segregation even though it was socially approved; thus, he pushed America to change for the better. In China, I added the example of Deng Xiaoping, who opposed widely approved Chinese thinking about how to run the economy; he instead tried to incorporate the best ideas from East and West—and this led to China's huge transformation over the last few decades.
- As an example of how the West can learn from the East, I appealed to how my illnesses were helped by a medical herb from the East (ginseng) which has been

[40] See http://harryhiker.com/china

[41] See https://www.zazzle.com/store/harrygensler. I'm now trying to bring out similar GR shirts for further languages and religions.

used for many centuries in China and has strong medical data in its favor, even though my Western doctors ignored it.

- I often apply ethical views to racism. I asked my students, "Were the Chinese ever victims of racism?" Yes, Europeans, thinking that they had a right to conquer other races, colonized parts of China in the nineteenth century; Wuhan itself was once an English colony. And Chinese people living in Western countries were often treated badly; Deng Xiaoping experienced this as a young foreign student in France.
- I often apply ethical views to global warming. Here I told the students that their country (China) and mine (America) are the two major players. We are the two great economic superpowers and the two great polluters; what we do will hugely impact the future of the planet. China, after a slow start, is making much progress; America, after a fair start, was moving backward (we're now doing better). We need to work together on this. We need a big war—not against each other, but against forces that are destroying the planet. If China and America can work together on this, then other countries may fall in line, and there's hope for the future of the planet.

After returning to the United States, I mentioned some of these points to my American students.

So what about the reaction papers from my Chinese students about cultural relativism? They were very much like the papers I get from my American students. Many students loved CR and praised its virtues. Some saw problems and tried to explain them. Some said that it all sounded good, but they were new to this and wanted to hear more. As in the United States, later reaction papers were hugely better and more reflective.

The best CR reaction paper was by my student April Rao 饶司立, a political science major who wants to be president of China someday. While she had never before heard of cultural relativism, her reaction in class was "Wow—that sounds reasonable and novel!" After thinking about it further, she found problems with CR. She decided that some basic values (like equality, democracy, rule by law, freedom, and independent thinking) are universal and should be accepted in every culture. These values are like cornerstones, or foundations, which support other cultural values (as added roofs and windows). Later in the course, April became a big supporter of the golden rule; her final paper was about how she'd someday run for president of China on a KITA golden rule platform.

References

Basler, R. (Ed.). (1953). *The collected works of Abraham Lincoln* (8 Vols.). Rutgers.

Benedict, R. (1934). A defense of cultural relativism. *Journal of General Psychology, 10*, 59–82.

Bloom, P. (2013). *Just babies*. Crown.

Darwin, C. (1859). *The origin of species*. John Murray.

Darwin, C. (1871). *The descent of man*. John Murray.

Festinger, L. (1957). *A theory of cognitive dissonance*. Stanford University Press.
Gensler Harry, J. (1996). *Formal ethics*. Routledge.
Gensler Harry, J. (2013). *Ethics and the golden rule*. Routledge.
Gensler Harry, J. (2016). *Ethics and religion*. Cambridge University Press.
Gensler Harry, J. (2017). *Introduction to logic* (3rd ed.). Routledge.
Gensler Harry, J. (2018). *Ethics: A contemporary introduction* (3rd ed.). Routledge.
Glynn, P. (1997). *God: The evidence*. Prima.
Hare, R. M. (1962). *Freedom and reason*. Oxford University Press.
Kant, I. (1964). *Groundwork of the metaphysics of morals* (H. J. Paton, Ed.). Harper & Row.
Kinnier, R., Kernes, J., & Dautheribes, T. (2000). A short list of universal moral values. *Counseling and Values, 45*, 4–16.
Kohlberg, L. (1972). A cognitive-developmental approach to moral education. *The Humanist, 32*, 13–16.
Kohlberg, L. (1981, 1984). *Essays on moral development* (Vols. 1 and 2). Harper & Row.
Mead, M. (Ed.). (1959). *An anthropologist at work: Writings of Ruth Benedict*. Houghton Mifflin.
Pfaff, D. (2007). *The neuroscience of fair play*. Dana.
Rimland, B. (1982). The altruism paradox. *Psychological Reports, 51*, 221–222.
Sazaklis, J., & Vecchio, L. (2015). *Wonder woman: An origin story*. Camstone.
Sheiman, B. (2009). *An atheist defends religion*. Penguin.

Harry J. Gensler was born in Detroit, Michigan, USA, in 1945. Gensler died at St. Camillus Jesuit Community in Wauwatosa on June 8, 2022. He received a doctorate in philosophy at the University of Michigan, writing a dissertation on "The Golden Rule" (roughly, "Treat others as you want to be treated"), under William K. Frankena. He is a Jesuit Catholic priest and taught for many years, mostly at Jesuit schools like Loyola University (Chicago) and John Carroll University (Cleveland), but also once at the University of Wuhan (China). He has published 16 books, mostly on logic and ethics; individual books have been translated into Arabic, French, Korean, Portuguese, Thai, Chinese (two books), and Persian (four books). He is arguably the world's expert on the golden rule, which is featured in several of his books. He is an avid hiker, having hiked the length of two of the longest trails in America: the Appalachian Trail, 2100 miles from Georgia to Maine, and the Pacific Crest Trail, 2600 miles from Mexico to Canada. He bicycled 3600 miles from Los Angeles to New York. And he is also very interested in computers, having produced specialized software for learning ethics and logic. See http://www.harryhiker.com for more information.

Clashes of Culture

Michael Smith

Abstract People from one cultural background often find themselves wondering how those from a different cultural background could willingly live the way they do and/or criticize others for living in the way they do. How are we to explain this kind of mutual incomprehension? Three different possible explanations will be considered. The first is that such mutual incomprehension is sourced in a moral flaw. The flaw might be in one of the cultures, as certain cultures might be premised on a moral mistake. Alternatively, the flaw might be a feature of the incomprehending individuals. If only they tried harder, they would come to appreciate the other's point of view. The second possibility is that such mutual incomprehension is the product of principled limits on the interpretation of someone from another culture. Contrary to the first possibility, there might be no way for people from different cultures to even understand each other, let alone find fault in each other, because the culturally mediated ways in which they understand the world are inaccessible to each other. The third possibility is that the mutual incomprehension is sourced in metaphysics. Contrary to the second possibility, there might be some non-metaphorical sense in which people from different cultures inhabit different worlds. The main task of the chapter is to describe and evaluate these three possibilities with a view to ruling out the third.

Keywords Carole Rovane · Positive ethics · Critical ethics · Disagreement · Ethical relativism · Constitutivism · Ethics of agency

Clashes of culture are ubiquitous. Here is an example that's close to home.

My mother was born in 1919 and grew up in a rural community in southeastern Australia. My father was born in 1918 and grew up in one of Melbourne's inner industrial suburbs. Because their families struggled, each of them had to leave school at the end of the eighth grade to help make ends meet. Honest and

M. Smith (✉)
Princeton University, Princeton, NJ, USA
e-mail: msmith@princeton.edu

S. Chakraborty (ed.), *Human Minds and Cultures*,
https://doi.org/10.1007/978-3-031-56448-2_4

self-reliant, they married in 1945, worked hard, and bought a house in the outer suburbs of Melbourne. As luck would have it, that suburb became the home of one of Melbourne's best government-funded high schools, a school whose students disproportionally went on to become artists, academics, scientists, health professionals, lawyers, engineers, and business leaders. Born there in the 1950s, my older brother and I enjoyed all the privileges enjoyed by the other baby boomers who went to our high school, much to the delight of our parents whose main hope was that our lives would be better than theirs. However, we also experienced many of the same clashes of culture with our parents that other baby boomers had with theirs.

By way of a window into our family, though we were the first on our street to get a TV, we had just one small bookshelf. The books I remember having are a dictionary; a set of Reader's Digest abridged versions of classic novels, versions that I have no memory of anyone reading; and *Vitalogy*, a compendium of homeopathic health remedies and advice on how to live a decent life that included a memorable chapter on the dangers of masturbation. After finishing high school, where his being popular with the girls had been a cause of great concern for my parents, my brother grew his hair, joined a country rock band as a drummer, and eventually moved out of home to live with his girlfriend. He slept most of the day, practiced with his band in the afternoons, and played at venues around Melbourne during the nights and early hours of the morning. I finished high school the year after my brother and went straight to university. Though I had many cousins, I was only the second member of our extended family to get a degree.

My brother and I were never quite sure what to make of our parents, especially our father, and, as you might expect, we had no real idea what they made of us. If we think of a culture are constituted by a set of shared expectations about the things that people do, say, and feel, then such mutual incomprehension is often present when there is a clash of cultures. In our case, the clash was multidimensional: part generational, part educational, part psycho-sexual, part aspirational, part aesthetic, and, once we entered the workforce, part socio-economic. But though we may not have known what to make of our parents, in the sense of knowing what it would be like to want to do and say many of the things that they said and did, we certainly knew why they were the way they were.

The education, aspirations, sexual attitudes, and aesthetic options available to my parents growing up during the Great Depression were limited in ways that ours weren't, and the choices that we made, given the options we had, inevitably resulted in not just attitudinal differences, including an attitude to risk that they thought imprudent, but rewards that were unavailable to them. Moreover, the awareness we had of their backgrounds meant that we also knew that, had we grown up in their circumstances, our attitudes, and lives too, might have been a lot like theirs. The imaginative obstacles we faced in thinking about what it would be like to be our parents were therefore in principle surmountable, and, in any case, we knew well enough to be able to empathize with them.

Notwithstanding the differences between us, in retrospect, there was another dimension along which our clash of cultures was less dramatic. My parents were loyal to their families, honest, and hard-working, and they wanted the very best for their children. Since I expect our friends would say much same about my brother and me, if they were asked what we are like, the clash of cultures that we experienced with our parents is therefore more accurately described as a clash of subcultures within a broader ethical culture—that is, a set of expectations about and understandings concerning what we owe to each other—that we shared with them, an ethical culture that largely persisted while other aspects of the culture changed dramatically.

This is not to say that there was no clash of ethical cultures with our parents. Though we agreed about the importance of loyalty, honesty, and self-reliance, we disagreed about the permissibility of pre-marital sex, and that pre-figured a larger disagreement about the roles men and women can and should play in society at large. To anticipate what is to come, in understanding these disagreements, we don't need to go beyond the positive ethics of the culture we shared with our parents: that is, the content of the ethical beliefs that members of a culture would have if they ordered their ethical beliefs into an internally coherent whole and then teased out all of their implications. Our parents had all sorts of beliefs about what was fair and what wasn't. They just hadn't thought through how their views about fairness undermined their views on relations between the sexes.

There are, however, other clashes of ethical culture where a shared positive ethic seems to be absent. In *The Metaphysics and Ethics of Relativism*, Carole Rovane tells the following story about one such clash (Rovane, 2013, pp. 41–44). (Rovane uses the first person in telling her story, but I will follow Kirk Ludwig in calling the protagonist of her story 'Anne' (Ludwig, 2015).) Anne is a middle-aged American woman who, through her own efforts at college and then her job on Wall Street, has ended up a contented, single, wealthy, urban retiree who spends her time doing pretty much whatever she pleases. Her parents, though proud of her achievements, would have much preferred her to get married and raise children somewhere near them in the Midwest, much as the other members of her family have done, but they recognize that her life choices were hers to make. Moreover, despite being much poorer than she is, they don't think that she should share her wealth with them. In their view, she deserves the money she has earned and is permitted to spend it on herself.

One day, while on holiday in India, Anne meets Anjali in a rural village in the Punjab, and the two women swap their life stories through an interpreter. They are much the same age, but Anjali is in a marriage that her parents arranged for her when she was very young and has spent her whole life, and continues to spend her life, raising the children she had with her husband and caring for the various members of her extended family, much as is the custom in her part of India. Whereas Anne is a contented, single, wealthy, and urban retiree who spends her time doing what she pleases, Anjali is a contented, married, poor, and rural woman who has always worked, and continues to work, long hours caring for others. The two women initially disapprove of each other's life choices.

Anne thinks that Anjali's parents had no right to arrange a marriage for her and supports this by talking about the value of self-determination. Anjali thinks that Anne wronged her parents by ignoring their wishes and supports this by talking about the duties associated with *kartavya*, that is, the special duties one has in virtue of one's place in one's cultural and social network. In Anjali's case, her special duties are a function of her role as a care-giver for the members of her extended family. These attempts to persuade the other are, however, unsuccessful. They come to accept that each other's choices make sense in their own circumstances, but they do so without thinking that the other's values provide them with a reason to rethink their own choices in their own circumstances, or any reason to think that if they were to change their circumstances, they should live by the other's values.

It is worth contrasting this case with the disagreement my brother and I had with our parents about the roles it is appropriate for men and women to play in society at large. Anne and Anjali evidently begin by thinking that their disagreement is like ours, that is, that it is a disagreement where a shared positive ethic lies in the background. There is a difference in ethical beliefs. Both parties to the disagreement initially think that that difference is the result of one or the other having an internally incoherent set of ethical beliefs, or of their having failed to tease out their ethical beliefs' implications. They therefore engage with each other with the aim of getting the other to see the error of their ways. But whereas my brother and I continue to think that this was the explanation for the disagreement with our parents, notwithstanding the fact that our father failed to change his mind, Rovane tells us that Anne and Anjali realize that this is not the explanation. The appropriate attitude for them to take to each other, as she sees things, is therefore one of disengagement. (In what follows, I will mostly assume that this version of Rovane's story is fully coherent and correct, but whether or not it is fully coherent, and hence correct, will come up at the very end.)

One reason Rovane gives for thinking that disengagement is the appropriate attitude for each of them to take has to do with the thick concepts that they use to express their ethical claims. When Anne talks about self-determination, Rovane thinks that she uses a concept that gets its significance from the experiences one has when one acquires the concept while living in a liberal culture like that of the American Midwest. When Anjali talks about kartavya, by contrast, she uses a concept that gets its significance from the experiences that one has when one acquires the concept while living in an extended-family-centred culture like that in rural India. Because the significance of the thick concepts they use when they make their ethical judgements is fixed by the experiences had when they acquired these concepts in their different cultures, and because neither Anne nor Anjali has enjoyed immersion in a culture that makes available the experiences required for possession of the other's concepts, Rovane thinks that neither Anne nor Anjali fully possesses the concepts the other does, and that neither is therefore in a position to fully understand the ethical judgements that the other makes.

If this is right, then Anne and Anjali's clash of ethical culture is indeed one in which there need be no shared positive ethic. If neither understands the ethical claim the other makes, then neither is in a position to entertain the other's ethical claim, let alone have a view about whether it would or would not survive if they ordered their ethical commitments into a coherent whole. Disengagement is the appropriate attitude for them to take to each other because there is nothing for them to engage with. When each of them grants that the other's choice makes sense in their circumstances, they therefore give each other the benefit of the doubt. Indeed, Rovane makes the further suggestion that the fact there is nothing for them to engage with may be a necessary feature of their disagreement, not just a contingent feature; that no one could enjoy cultural immersion of the kind required to possess both concepts. Possessing the thick concept of self-determination may make it impossible to possess the thick concept of kartavya and vice versa.

What should we make of this further suggestion? Think again about the imaginative obstacles my brother and I faced in knowing what it would be like to want to do and say many of the things that our parents said and did, given how very different our lives were from theirs. As noted earlier, merely knowing that our parents were the way they were because of their backgrounds made it clear to my brother and me that, had we grown up with backgrounds like theirs, then our inner lives might well have been just like theirs. The imaginative obstacles we faced in knowing what it would be like to be our parents were therefore contingent, not necessary, because they are absent altogether in the possible world in which we grow up having a qualitatively identical background to theirs. Moreover, they are only present in the actual world to the extent that we have difficulty imagining that possible world. Contrary to Rovane's further suggestion, reflection on this case provides us with a recipe for constructing a possible world in which Anne and Anjali overcome the obstacles they face in understanding each other.

Consider a possible world in which (say) Anne first of all lives her own life from birth to death, thereby acquiring the concept of self-determination via immersion in a liberal culture like that in the American Midwest; is then reborn and lives a life qualitatively identical to Anjali's from birth to death, thereby acquiring the concept of kartavya via immersion in an extended-family-centred culture like that in rural India; and is then reborn again and remembers each of the past lives she led in full detail. If this description of the world we are to consider is coherent, as it at least seems to be, then in that possible world, Anne meets all of the conditions for fully understanding not just her own ethical claims but also those of Anjali. If this is right, then such understanding is not metaphysically impossible. Of course, it may still be a physical impossibility for anyone in the actual world to fully understand both claims, as it may be impossible for someone to fully imagine that possible world, so the clash of cultures may in that sense still be insurmountable. Crucially, however, there seems to be no in-principle insurmountability.

Rovane offers another reason for thinking that Anne and Anjali's clash of cultures is one in which there is no shared positive ethic in the background, and that the appropriate attitude for them to take to each other is therefore one of disengagement. The reason is premised on her relativism about ethical claims: that is, Rovane's view that ethical claims are always made in a culture that gives those claims their significance, and that they are therefore made true by facts about that culture. Though she seems to think that there is an important connection between her ethical relativism and the claims about the difficulty people from different cultures have understanding each other's ethical claims, her ethical relativism is in fact logically independent of the difficulty or ease with which people from different cultures grasp each other's ethical claims. Indeed, Rovane gives at least one reason for being an ethical relativist that has nothing to do with what it takes to understand ethical claims. When we evaluate the truth of an ethical claim, she suggests that we do so by seeing whether it is invulnerable to a decisive local objection, and, she seems to think, it follows from this that being invulnerable to a decisive local objection is the truth-maker of that ethical claim (Rovane, 2015).

Focus on Anne. The idea is that in evaluating the truth of her claim that she was within her rights to lead the life of a single urban professional, notwithstanding the fact that her parents would have preferred that she had married and lived nearby, we are to ask ourselves the following question. Is that claim is invulnerable to a decisive objection on the basis of the full suite of ethical commitments made available to her by growing up in a liberal culture like that in the American Midwest, once these commitments have been ordered into an internally coherent whole and their implications teased out? This is equivalent to Anne's claim's being entailed by the positive ethics of a liberal culture, in the sense of 'positive ethics' defined earlier. According to this way of thinking about the truth-maker of Anne's ethical claim, Anjali's objections are completely irrelevant to the truth of her claim whether she understands Anne's ethical claim or not. The same goes for Anne's objections to the truth of Anjali's ethical claim. What makes Anjali's claim that her parents are within their rights to arrange a marriage for her true is the fact that that accords with the positive ethics of an extended-family-centred culture. The upshot is that even if Anne and Anjali do understand each other's ethical claims, their objections, grounded as they are in their own culture, are irrelevant to the truth of the other.

Why is disengagement the appropriate attitude for Anne and Anjali to take to each other if this is the situation in which they find themselves? Disengagement is the appropriate attitude for them to take because they have nothing to offer each other by way of objection unless they understand each other's culture and concepts well enough to be able to tell whether there are any decisive objections to the ethical claims the other makes based on the full suite of ethical commitments made available to the other by growing up in the other's culture. Rovane's view is, of course, that neither Anne nor Anjali meet this condition, and that perhaps no one could. Having said that, it is important to note that disengagement of the kind Rovane recommends need not entail disengagement in the sense of tolerating the other's acting in accordance with their positive ethics. Whether they should tolerate the other's acting is, after all, an ethical claim, so Rovane's ethical relativism tells us that the truth of that claim depends on whether such toleration is invulnerable to a local objection. Disengagement in the sense of giving up trying to find a local objection to another's ethical claim may therefore be consistent with practical opposition. (For more on this, see Bernard Williams's discussion of the 'anthropologist's heresy' (Williams, 1972).)

All of this is, however, premised on Rovane's ethical relativism. But should we agree that the truth of an ethical claim made by someone from a certain culture is a matter of the invulnerability of that ethical claim to a decisive local objection? Are the only objections we can make to an ethical claim based on the full suite of ethical commitments made available to the person who makes that claim by growing up in their culture once these commitments have been ordered into an internally coherent whole and their implications teased out? The answer to both of these questions is an emphatic no. To see that this is so, we need to think about three related points: the first is the ways in which someone who has ethical obligations can think of themselves that are not tied to their being located in their culture, the second is what their capacity to think of themselves in these ways tells us about the reasons for action they have, and the third is the relationship between reasons for action and ethical truths.

Consider the first. Everyone who has ethical obligations can think of themselves as a distinctive kind of object in the spatio-temporal order. One thing that distinguishes the kind of object they are from objects of other kinds in the spatio-temporal order is the fact that they have both desires about the way the world is to be and beliefs about the things that they can do so as to make the world that way. In virtue of having such desires and beliefs, they are therefore one of the many objects in the spatio-temporal order that can act: that is to say, they can make the world the way they desire it to be, so satisfying their desires. In a word, all agents who have ethical obligations can therefore think of themselves as *agents*. But though all of those who have ethical obligations are agents, and can think of themselves as such, not all agents have ethical obligations. Examples of agents who lack ethical obligations include non-human animals, human infants, and some severely cognitively impaired mature humans. (We will have more to say about what our ethical obligations are in due course, including what our obligations are to agents who have no ethical obligations themselves.)

Now consider the second point. What does the capacity that those with ethical obligations have to think of themselves as agents tell us about the reasons for action that they have? Since they can think of themselves as agents, it follows immediately that they can think of themselves as having *motivating* reasons for action, as these are those psychological states that cause and rationalize their behaviour (Smith, 1994). They can think of themselves as having motivating reasons because they can think of themselves as having the particular desires and beliefs that they have, the ones that will in fact cause and rationalize their behaviour when they act. But there is also a different sense of the term 'reason' in which agents have reasons for action, a normative sense, and those with ethical obligations are also in a position to think of themselves as having reasons for action in this sense. Agents' *normative* reasons for action are those considerations that would justify their acting in a certain way, whether or not they are motivated to act in that way (Smith, 1994). Those with ethical obligations are in a position to think of themselves as having normative reasons for action because the concept of a normative reason for action can itself be understood entirely in terms of the concept of agency. Here is how.

In thinking of themselves as agents, those who have ethical obligations think of themselves as a member of what Judith Jarvis Thomson calls a goodness-fixing kind (Thomson, 2008). The kind *agent* is a goodness-fixing kind because all agents can be placed somewhere along an ordering from best to worst according to how well they measure up to a standard that is internal to agency itself. What is this standard? As we have already seen, in virtue of having desires and beliefs, agents can act in the world: that is to say, they can do what they have motivating reasons to do, so making the world the way that they desire it to be, thereby satisfying their desires. But though they can satisfy their desires, they may not; even if they do satisfy their desires, they may not do so reliably, and even when they do reliably satisfy their desires, this may still be the result of a happy accident of a different kind.

For example, those who act on false beliefs about how they can fail to satisfy their desires will almost always fail to satisfy their desires, and those who act on beliefs that are true, but only accidently true, will only satisfy their desires because of that happy accident. The former are therefore located somewhere low down on the ordering of agents from best to worst, the latter above them but lower down than agents who not only satisfy their desires but also do so in a reliable fashion. These last agents, those who reliably satisfy their desires, don't just have true beliefs about how to satisfy their desires. They have knowledge of how to satisfy them. But agents who know how to satisfy their desires, but only because their desires are limited to ways the world can be that they happen to have the capacity to know how to bring about, satisfy their desires merely because of a happy accident as well. They are therefore ranked lower down the ordering of agents from best to worst than agents whose know-how isn't limited in this way. These are agents who know how to satisfy their desires and would still know how to satisfy them even if they had different desires.

As we move up the ordering of agents from best to worst, we therefore minimize the extent to which accidents play a role in their having knowledge and satisfying their desires. The upshot is that at the very top of this ordering, we find those for whom the satisfaction of their desires is no accident at all. These agents—let's call them *ideal* agents—robustly have and exercise the capacity to satisfy their desires, no matter what they desire, and they do so because they robustly have and exercise the capacity to know what the world is like, no matter what it is like. An important consequence of this is that, as we move down the ordering of agents from best to worst, those who are non-ideal, but only a little non-ideal, still do exceptionally well in knowledge-acquisition and desire-realization terms. Their vulnerability to doing badly in the face of changes in their contingent circumstances is therefore minimal.

I said earlier that the concept of a normative reason for action can be understood in terms of the concept of agency. Consider an agent who has certain desires and beliefs about how to realize those desires. He has a motivating reason to act accordingly in those circumstances. But does he have a normative reason to act accordingly? Would he be justified in doing so? Not necessarily. He has a normative reason to act in that way in those circumstances only if his ideal counterpart—that is, himself in the nearest possible world in which he robustly possesses and exercises maximal capacities for desire-realization and knowledge-acquisition—desires that he acts in that way in those circumstances. An agent's ideal counterpart has an action plan for every circumstance in which he could possibly find himself, and that includes the circumstance in which the non-ideal agent finds himself. When a non-ideal agent acts, he therefore either does what his ideal counterpart wants him to do in those circumstances or fails to act in that way, and what he should do is act in that way. The normativity of reasons for action—that is, the sense in which there are considerations that justify acting in certain ways rather than others—is thus connected to an action's being one that the best kind of agent would want to perform, in a sense of 'best' that is internal to agency itself.

At this point, I can imagine that the ethical relativist is beginning to grow impatient. 'Sure enough', she will say, 'someone who has ethical obligations has normative reasons for action, but these normative reasons will typically coincide with the positive ethics of her culture, and hence with the ethical truths in that culture. Think of what happens when children grow up in an ethical culture. Their parents typically cause them to acquire not just beliefs about the ethical obligations they have within that ethical culture but also desires to act accordingly. Some of the ethical beliefs they lead them to acquire will be decomposable into ethical and non-ethical component beliefs, as there are both fundamental ethical beliefs—that is, beliefs about fundamental values or end-states—and non-fundamental ethical beliefs—that is, beliefs about the means that need to be taken to realize some fundamental value or end-state. This mirrors a distinction between the desires that parents lead their children to acquire. Some of these are desires for end-states, and some are desires for means to those end-states, which are properly understood as wholes that have desires for end-states and beliefs about means as parts. Parents can, of course, cause their children to have false ethical beliefs, as some might not survive once the full suite of ethical beliefs the children are led to acquire within that culture is ordered into a coherent whole and the implications of these beliefs are teased out. This too parallels a way in which the desires that parents cause their children to have may fail to coincide with the desires that children's ideal counterparts have. The various desires for end-states they cause them to have might not cohere with each other, the beliefs about means to ends they cause them to have might be false, and they may fail to cause them to have relevant beliefs about means. So though it is true that there are normative reasons in addition to ethical obligations, in the typical case these are acquired in parallel. The upshot is that normative reasons are not an independent source of local objection to any ethical claims, or at any rate they aren't in the typical case'.

The ethical relativist's impatience makes clear the crucial point of disagreement. The ethical relativist thinks that whatever desires for end-states agents have, whether those agents are ideal or non-ideal, they have them in virtue of some factor exogenous to their being agents. In the typical case, they have the desires for end-states they have in virtue of being enculturated by their parents. Being well brought up in a culture is a matter of being made to desire to do what they are ethically obliged to do. Since these desires will be shared by their ideal counterparts, they are therefore enculturated to have normative reasons to do what they're ethically obliged to do. But the mere fact that they are agents places no constraints at all on the desires for end-states that they have. Badly brought-up agents will be raised to desire to act in ways that violate their ethical obligations. That is what they will have normative reason to do. So, at any rate, the ethical relativist thinks.

But on reflection, it seems that the mere fact that ideal agents are agents does place constraints on the desires for end-states that they have, and it further seems that this, together with the fact that some non-ideal agents have the capacity for self-control, where this is a capacity to get yourself to desire to do and do what you have normative reason to do, means that the fact that they are agents explains why some non-ideal agents have at least the some of the desires that they have as well (what follows draws on Smith, 2021). The argument for this conclusion begins from the premise that, as have already seen, ideal agents robustly possess and exercise maximal capacities for desire-realization and knowledge-acquisition (let's call these their agential capacities). Given that agents are temporally extended, it follows from this that the robust possession and exercise of their agential capacities must itself be temporally extended, as the robustness of an agent's possession and exercise of their agential capacities is a measure of how well they are doing in terms of the possession and exercise of those capacities at each moment they exist. Ideal agents must therefore have the wherewithal within themselves, or the worlds in which they exist contain the wherewithal, to overcome two kinds of vulnerability to which they would otherwise be subject.

The first kind of vulnerability is the vulnerability of an agent's later self to their earlier self. Imagine an agent at a certain time who exercises their agential capacities at that time. For their possession and exercise of this capacity to be robust, it cannot be the case that that exercise is dependent on the fact that, at an earlier time, they just so happened not to want to deceive themselves at the later time, or that at the earlier time, they just so happened not to want to undermine their satisfaction of their desires at the later time. This would make their being ideal the result of a happy accident. Moreover, if their possession of those capacities is itself the product of their efforts to develop and maintain their capacities at some earlier time, then it couldn't be dependent on the fact that, at that earlier time, they just so happened to want to help make sure that they develop and maintain such capacities to exercise at a later time. This too would make their being ideal the result of a happy accident. They must instead have some feature, in virtue of being an agent who possesses and exercises all of the capacities that underwrite their being an agent at each moment they exist robustly, that explains why at the earlier time they helped develop and maintain their having their agential capacities, and why at the earlier time they didn't interfere with their later exercise.

The obvious feature to imagine them having at the earlier time to overcome this vulnerability of their later self to their earlier self is a pair of desires for end-states, constitutive of being an agent whose exercise of their agential capacities is robust. These are a desire to help ensure that that they have agential capacities to exercise, on the one hand, and a desire not to interfere with their exercise of these capacities once they have them on condition that their exercise won't itself constitute interference, on the other. Another obvious feature to imagine them having is the self-control required to make sure that they act in accordance with such desires in nearby worlds in which they are subject to apathy or weakness of will. Agents who lack these two desires, or who lack the capacity for self-control, are thus, in virtue of that very fact, less than ideal.

The second kind of vulnerability is the interpersonal analogue of the intrapersonal vulnerability just described. Imagine an agent at a certain time who possesses and exercises the capacity to know what the world is like and satisfy their desires in the world at that time. For their possession and exercise of this capacity to be robust, it cannot be the case that that possession and exercise is dependent on the fact that, at that time, there are other agents who just so happen not to want to deceive them, or that at that time, there are other agents who just so happen not to want to undermine their satisfaction of their desires. This too would make their being ideal the result of a happy accident. Moreover, if their possession of those capacities at that time is the product of the efforts of other agents to help them develop and maintain their capacities, it couldn't be dependent on the fact that those other agents just so happened to want to help them develop and maintain their capacities. This would make their being ideal the result of a happy accident as well. Those other agents must instead have some further feature that explains why they help but don't interfere.

What this suggests is that we would be wrong if we were to suppose that agents could robustly exercise their agential capacities in worlds in which other agents have just any old desires for end-states, and that we would be wrong to restrict the desires that agents have to overcome their intrapersonal vulnerability to desires that concern themselves. For an agent's exercise of their agential capacities to be robust, they must instead be in the company of other agents who robustly exercise their agential capacities too, and, in virtue of their being ideal, all ideal agents must have a pair of desires. These are, first, a desire to help develop and maintain the agential capacities of every agent, themselves and other agents as well, and second, a desire not to interfere with the exercise of the agential capacities of any agent, themselves and others too, on condition that exercise wouldn't itself constitute interference. For short, let's call these the desires to help but not interfere. Moreover, all such agents must have the capacity for self-control required to make sure that, in nearby worlds in which they are subject to apathy or weakness of will, they act in accordance with such desires anyway. Agents who lack the desires to help but not interfere or the capacity for self-control are thus less than ideal in virtue of that very fact.

We saw earlier that an important consequence of the fact that ideal agents robustly possess and exercise their agential capacities is that, as we move down the ordering of agents from best to worst, those who are non-ideal, but only a little non-ideal, are still doing exceptionally well in terms of knowledge-acquisition and desire-realization. Their vulnerability to doing badly in the face of changes in their contingent circumstances is minimal. We can now see why this is so. Their vulnerability is minimal because ideal agents themselves have desires to help everyone but not interfere with anyone, themselves included, so in nearby non-ideal possible worlds to the ideal world in which they are not disposed to help but not interfere, they have the self-control required to get themselves to do so anyway. Moreover, since they are surrounded by other agents who are themselves ideal, their vulnerability to not being helped or to being interfered with by them is minimized for the same reason. In possible worlds nearby the ideal world in which those they are surrounded by are not disposed to help them, and are disposed to interfere with them, those that surround them also have the self-control required to get themselves to help but not interfere anyway.

When we put this account of what ideal agents are like together with the earlier account of what it is for an agent to have a normative reason in certain circumstances in terms of what their ideal counterpart desires that they do in those circumstances, we get the following result. Everyone, no matter what their non-ideal circumstances, has three kinds of normative reason to act in those non-ideal circumstances. The first is a normative reason to satisfy whatever desires they happen find themselves with in their non-ideal circumstances. This is because, since their ideal counterpart is exactly like them, except for being ideal, their ideal counterpart has all the same desires that they have concerning their non-ideal circumstances. These desires are indeed caused in agents by factors exogenous to their being agents, and the cause with which we're most familiar is enculturation. The second is a normative reason to help make sure that everyone in their non-ideal circumstances has capacities for knowledge-acquisition and desire-realization to exercise. Every agent's ideal counterpart has that desire in virtue of being ideal. The third is a normative reason not to interfere with anyone in their non-ideal circumstances' exercise of the capacities for knowledge-acquisition or desire-realization. Every agent's ideal counterpart has that desire in virtue of being ideal as well. Put more simply, every agent, no matter what their non-ideal circumstances, has normative reasons to help but not interfere, and, apart from that, they have a normative reason to do whatever they want to do in those circumstances.

Remember the crucial point of disagreement with the ethical relativist. She thinks that agents have all the desires they have in virtue of factors exogenous to their being agents. Those who are well brought up within a culture acquire desires to do what they are ethically obliged to do, relative to that culture, and since these desires will be shared by their ideal counterparts, those who are well brought up within a culture will have normative reasons to fulfil their ethical obligations. But those who are badly brought up acquire desires to act in ways that are contrary to their ethical obligations, and such agents therefore have normative reasons to act in those ways. But we can now see that this view of the relationship between normative reasons and ethical obligations is mistaken and quite radically so. Every agent, no matter how they are brought up, has a normative reason to help, a normative reason not to interfere, and a normative reason to do whatever they want to do. Since the normative reasons to help but not interfere specify what agents owe to each other, simply in virtue of being agents, these two normative reasons have distinctively ethical content. The normative reason all agents have to do whatever they desire to do, by contrast, has non-ethical content. With this understood, four further points about these normative reasons, and about the relationship between them and ethical obligations, permissions, and prohibitions, need to be borne in mind.

The first is that there is indeterminacy in the content of the two ethical reasons. Consider the reason not to interfere with anyone's exercise of their capacities for knowledge-acquisition and desire-realization, by way of example—similar points could be made about the reason to help. If you cause someone to be in extreme agony, then they may be so overwhelmed that they are unable to think or act at all. But now imagine a continuous sliding scale of pains with extreme agony at one end and only the faintest feeling of pain at the other, a feeling so faint that it doesn't interfere with anyone's exercise of their agential capacities, and then imagine a range of actions that cause the pains that lie at each point along this sliding scale. At what point do these actions amount to interference with an agent's exercise of their capacity to know what the world is like and realize their desires in it? The answer is that there is no such point. Instead, there is a range of actions that are determinately instances of interference and a range of actions that are determinately not instances of interference, and in between there is a range of cases about which it is indeterminate whether they are cases of interference.

The second point concerns the weights of the three reasons. To stick with the example of the reason not to interfere—similar points could be made about the other two reasons—this reason will be more or less weighty depending on the extent to which the interference in question undermines an agent's exercise of their agential capacities. Given that the three reasons can conflict with each other, this means that there is no easy way to spell out the relative weights of the three reasons in cases where they conflict. To return to the simple way of spelling out the content of the three reasons, the option best supported by all three reasons taken together will be that option in which the mix of doing what one wants, not interfering with anyone's exercise of their agential capacities, and making sure that everyone has agential capacities to exercise, is that which is most similar to the mix in the ideal case. In some cases, the ethical reason to help will outweigh the ethical reason not to interfere, in other cases vice versa. In some cases, the ethical reasons to help but not interfere will outweigh the non-ethical reason to do what one wants, and in other cases vice versa. Moreover, since there is bound to be vagueness in this similarity relation, the weights that these reasons have will themselves almost certainly be vague too.

The third point—and this is also the final point that needs to be made in order to see why we shouldn't be ethical relativists—is that, keeping in mind what's just been said about the vagueness and weights of the three reasons, we can now provide the following plausible accounts of what it is for actions to be ethically obligatory, permissible, and forbidden. Ethically obligatory actions are those that there are decisive ethical reasons to perform; ethically permissible actions are those that there are no decisive ethical reasons not to perform; and ethically forbidden actions are those that there are decisive ethical reasons not to perform.

With this account of ethical obligations, permissions, and prohibitions in hand, we can also clarify their scope. The class of agents is the class of beings who can act: that is, the class of beings with the capacity to realize their desires in the light of their beliefs about the world in which they live. This includes not just human animals but also non-human animals, and it includes not just able-bodied human and non-human animals in the prime of their lives but also newly born and infant human and non-human animals, and disabled human and non-human animals, including those in a state of physical and cognitive decline towards the end of their lives. There are therefore two issues about the scope of our ethical obligations, permissions, and prohibitions that require clarification. One is who has ethical obligations. The other is to whom they have those ethical obligations. Do all agents have ethical obligations? And do those who have ethical obligations have them to all agents? Here it seems to me we find an asymmetry.

Since the fact that agents have normative reasons to help but not interfere ensures that agents in non-ideal worlds fare well in knowledge-acquisition and desire-realization terms, all agents must fall within their scope. They must be normative reasons to help *every* agent and not interfere with *any* agent. But for an agent to have normative reasons to act in a certain way, that agent must at the very least have the capacity to think of itself as having the option to act in that way. This means that, for example, only beings who can think of themselves as having the option to help but not interfere have normative reasons to help but not interfere. Since not all agents have that capacity, not all agents have reasons to help but not interfere, and so not all agents are subject to ethical obligations, permissions, and prohibitions. Examples of such agents include newly born and infant humans, human animals in the final stages of physical and cognitive decline towards the end of their lives, and (to the best of our knowledge) all non-human animals.

Let's return to the disagreement with the ethical relativist. Note how very different this account of our ethical obligations is from hers. The ethical relativist thinks that the truth of an ethical claim made by someone in a certain culture is a matter of that claim's invulnerability to a decisive objection made on the basis of the full suite of ethical commitments made available to her by growing up in that culture. This in turn means that there is no distinction to be drawn between the ethical truth in a culture and the positive ethics of that culture. But as we have seen, and with a qualification to be mentioned presently, such a distinction can be drawn because we can object to the positive ethics of a culture on the basis of considerations that are independent of any culture. That's what the critical ethics of normative reasons allows us to do.

Importantly, this is not to deny that the culture in which an action is performed is a crucial determinant of what there are normative reasons to do in that culture. If cultures are constituted by a set of shared expectations about the things that people do, say, and feel, then to the extent we hold ourselves to such expectations, and feel comfortable around others who hold themselves to such expectations as well, cultures can therefore be understood as the source of many of the desires we have about the lives we wish to live. In this way, they help fix the content of many of the non-ethical desires we have normative reasons to realize, and, since ethical reasons have to outweigh non-ethical reasons in order for us to have an ethical obligation, they also help determine our ethical obligations. Moreover, the vagueness of the ethical reasons that we have independently of any culture means that there is a legitimate role for culture to play in giving these ethical reasons more determinate content. To the extent that our culturally specific expectations concern what it is to help others, and what it is to interfere with them, cultures are therefore among the determinants of the vague content of the ethical reasons to help but not interfere. Cultures should, to that extent, be maintained.

However, to the extent that these culturally specific expectations undermine our helping everyone and not interfering with anyone, the normative reasons we have to help but not interfere, and apart from that to do whatever we wish, provide us with an important insight into the ways in which the cultures in which we find ourselves need to be changed. For the reasons given, changing them is, however, a delicate balancing act of conflicting ethical and non-ethical reasons. Whether cultural expectations should be maintained or changed need not be an all-or-nothing matter. To the extent that a culture gives determinate content to helping and not interfering, those cultural expectations need to be maintained, but to the extent that they undermine helping and not interfering, they need to be changed. These changes should in turn give due weight to the non-ethical desires we may have to maintain our culture. It should be clear that undertaking a delicate balancing act of this kind would require considerable understanding of both the positive ethics of a culture and the critical ethics of agency.

In light of all this, what should we make of Anne and Anjali's clash of cultures? We saw earlier that Anne and Anjali begin by thinking that the difference in their ethical beliefs is the result of one or the other having a set of ethical beliefs that is internally incoherent, or of their having failed to tease out their ethical belief's implications, and they engage with each other with the aim of getting the other to see the error of their ways. Rovane tells us that Anne and Anjali are unable to identify any incoherence in each other's beliefs, or anywhere that the other has failed to tease out the implications of their ethical beliefs, and that they therefore disengage with each other in the sense of giving up that aim. She further explains this disengagement in terms of Anne and Anjali's coming to see that both of their beliefs are true, but that they are made true by different facts. Ethical relativism is thus the crucial explanation of their disengagement. But the critical ethics of agency suggests a very different diagnosis of Anne and Anjali's clash of cultures.

An initial point of contention concerns Rovane's claim that Anne and Anjali are unable to identify any incoherence in each other's beliefs, or anywhere that the other has failed to tease out the implications of their ethical beliefs. Since the ethics of agency is knowable by anyone capable of understanding what it is to be an agent, what the connection between agency and normative reasons for action is, and what the connection between normative reasons for action and ethical obligations is, it seems that any ethical culture that encourages that kind of critical thinking makes the critical ethics of agency available to its members. Since the story Rovane tells about Anne and Anjali doesn't specify whether their respective cultures do or do not encourage that kind of critical thinking, it is therefore unclear whether she is right to say that Anne and Anjali are unable to identify any incoherence at all in each other's ethical beliefs, or anywhere that the other has failed to tease out the implications of their ethical beliefs. If (say) Anne's culture didn't encourage that kind of critical thinking, but Anjali's did, then there is a great deal left for Anjali to say to Anne about the relevance of the critical ethics of agency to her ethical beliefs.

But putting that point to one side, the critical ethics of agency suggests that even if the critical ethics of agency were available to both Anne and Anjali, and even if they had thought hard about what the critical ethics of agency requires of them, they might still disengage with each other. For there is nothing in the story as told by Rovane that tells us whether Anne's belief that Anjali's parents had no right to arrange a marriage for her, or Anjali's belief that Anne wronged her parents by ignoring their wish that she marry, have children, and live nearby, aren't themselves true, according to the critical ethics of agency. It all depends on what we discover once we fully specify what their ethical reasons to help but not interfere, and apart from that do whatever they desire to do, support their doing in the context of their respective cultures. All of the important questions about the extent to which their respective cultures provide them with non-ethical reasons for action, and give determinate content within their culture to their ethical reasons for action, and hence the all-important question of whether they have decisive ethical reasons for maintaining or for changing these particular expectations within their culture, are left unanswered by Rovane's story as told. Moreover, given the difficulties involved in giving answers to these questions—here I have in mind both the difficulty, emphasized by Rovane, involved in grasping another's concepts, and also the delicate act of balancing ethical and non-ethical reasons referred to above—it isn't clear that in an exchange like that between Anne and Anjali, either of them would be in a position to claim an advantage over the other as to how these questions are to be answered in the other's culture.

The upshot is that, for all that we've been told, Anne may be right and Anjali wrong, Anne wrong and Anjali right, Anne and Anjali both right, and Anne and Anjali both wrong. Moreover, this is so notwithstanding the fact that both Anne and Anjali are bound by the critical ethics of agency, where the critical ethics of agency isn't itself grounded in any cultural facts at all. Disengagement with the other, motivated by humility about being the other's epistemic peer, may well therefore be the right attitude for Anne and Anjali to take to each other.

References

Ludwig, K. (2015). The sources of relativism. *Ethics, 126*, 175–195.

Rovane, C. (2013). *The metaphysics and ethics of relativism*. Harvard University Press.

Rovane, C. (2015, November 23). Moral dispute or cultural difference. *The New York Times*.

Smith, M. (1994). *The moral problem*. Wiley-Blackwell.

Smith, M. (2021). The state of nature. In M. Timmons (Ed.), *Oxford studies in normative ethics* (pp. 270–289). Oxford University Press.

Thomson, J. (2008). *Normativity*. Open Court.

Williams, B. (1972). *Morality: An introduction to ethics*. Cambridge University Press.

Michael Smith is currently McCosh Professor of Philosophy at Princeton University and Adjunct Professor of Philosophy at Monash University, having previously held teaching positions at Oxford, Monash, and The Australian National University. He is the author of *The Moral Problem* (1994) and *Ethics and the A Priori: Selected Essays on Moral Psychology and Metaethics* (2004) and the co-author with Frank Jackson and Philip Pettit of *Mind, Morality, and Explanation: Selected Collaborations* (2005). Smith gave the John Locke Lectures in Oxford in 2017, and the book of the lectures, *A Standard of Judgement*, will be published in due course by Oxford University Press.

Hatred: Why Do Such Nice People Do Such Awful Things?

Michael Ruse

Abstract Humans are by nature social. And yet, we humans can be so cruel to each other. The dreadful wars of the last century: the First World War, the Second World War, the Korean War, the Vietnam War, and so the list expands. Then there is the prejudice that members of one group show to members of other groups. Americans and slavery come at once to mind. So how do we explain the paradox? Why do such nice people do such awful things? I am an evolutionist, so I believe that the answers to the present are to be found in the past. Simply put, we humans were hunter-gatherers, and natural selection made us highly suited to our lifestyle. Then came agriculture and all changed. Our formerly efficient adaptations were too often not adequate for our new circumstances and lifestyles. Conflicts, group and individual, arose. I shall ask whether we can reconcile the tensions in our position. Can we move forward, bringing the knowledge of our biological past combined with the awareness of our cultural present to speak positively and creatively to the challenges that lie before us? I am ever an optimist. Is my optimism justified?

Keywords Hate · Ingroup/outgroup · Natural selection · War · Agriculture · Prejudice · Crowding

I have a book, *Why We Hate: Understanding the Roots of Human Conflict*, published by Oxford University Press, appearing at the same time as this chapter. I think it is one of the most important works I have written, discussing social issues with great impact on modern societies. Knowing only too well that today, thanks to electronic means of communicating, fewer and fewer are willing to read even relatively short books, as is mine, I am taking the opportunity to present a brief péecis, this chapter. It is probably overestimating the chances that readers of this chapter will be stimulated to read the whole book, but pursuing the impossible is what we philosophers do for a living.

M. Ruse (✉)
Florida State University, Tallahassee, FL, USA
e-mail: mruse@fsu.edu

S. Chakraborty (ed.), *Human Minds and Cultures*,
https://doi.org/10.1007/978-3-031-56448-2_5

Dusti Talavera was inside her Denver apartment on Sunday afternoon, watching a group of children play on a frozen pond outside her window, when suddenly the icy surface cracked. Then came a splash. Three of the children plunged into the frigid water.

"Before I even realized it, I was out there on the middle of the pond, pulling two kids out," Talavera, 23, said at a news conference on Monday.

But as she reached for the third child, the ice gave way beneath her, and Talavera, too, found herself in the 15-ft-deep pond. She tried to pull herself out, but the ice was too slick, so she treaded water to stay afloat as she struggled to hold the unconscious girl's head above the surface, she said.

The little girl's teenage cousin soon appeared, tossing Talavera a rope. She managed to get back onto the surface and pull the 6-year-old to safety (*The Washington Post*, January 11, 2022).

Read that without tears coming to your eyes. A young woman, with her whole adult life before her, without hesitation puts herself at serious risk to save the lives of three children—total strangers. Exceptional; except, not really. We live in groups, we get on with family—partners, parents, children—we interact harmoniously, and so much more. I have just retired after teaching philosophy to other people's kids for 55 years. I am not Mother Teresa. I have done what I did because I enjoyed it immensely. Not all students are warm and friendly, intelligent, wanting to learn. But the vast majority are, and those that are not are the challenge of the job. It is what makes it all so worthwhile, reaching out to those who are not easy. To use a trite but true saying, "The sale starts when the customer says 'No!'" That's the one side of us. Think of the sacrifices so many have made for us during the Covid pandemic—doctors, nurses, and lower-level hospital staff. Not our relatives, not our friends, probably strangers—but there to do their duty, which they have not shirked. My son, of whom I am immensely proud, is an emergency medic, out there comforting people and taking the old and sick to places where they can get the help they need. They are saints but not atypical. Humans do step up to the plate when the need arises.

Killer Apes

And yet, in his *The Better Angels of our Nature: Why Violence Has Declined,* best-selling, academic superstar Steven Pinker tells it like it is. We humans are a violent folk—"most of us—including you, dear reader—are wired for violence, even if in all likelihood we will never have an occasion to use it" (Pinker, 2011, p. 483). Although it is true that "when men confront each other in face-to-face conflict, they often exercise restraint. But this reticence is not a sign that humans are gentle and compassionate." Pure and simple. We are nice only out of caution or prudence. Pinker is part of a tradition that flowered in the years after the Second World War. Human beings are "killer apes." Everyone knew this from the primate antics at the beginning of the movie *2001*, but it was well established before this. The poster from The First World War shows that the trope goes back at least a century if not more, but it really sprang to life—to use an appropriate metaphor—in 1953.

Paleoanthropologist Raymond Dart was its modern originator. Rightly celebrated as the person who in 1924 identified Taung Baby, *Australopithecus africanus*—a human ancestor ("hominin")—by the 1950s Dart, who had encountered prejudice and disbelief for over 30 years, had no hesitation about venturing into speculation. To use another appropriate metaphor, he made no bones about his views. We were flesh eaters. And we were not that fussy about whom we ate. The appalling Jeffrey Dahmer murdered, many young men, and used to have (gay) sex after they were dead. He would then settle down for a nice meal of chunks of his victims and French fries. He was not as abnormal as one might think. In the good (?) old days, we did it all the time. Whether by choice or need, we ate each other. That explains a lot: "The loathsome cruelty of mankind to man forms one of his inescapable characteristics and differentiative features; and it is explicable only in terms of his carnivorous, and cannibalistic origin" (Dart, 1953). Dart was just warming to his theme:

> The blood-bespattered, slaughter-gutted archives of human history from the earliest Egyptian and Sumerian records to the most recent atrocities of the Second World War accord with early universal cannibalism, with animal and human sacrificial practices of their substitutes in formalized religions and with the world-wide scalping, head-hunting, body-mutilating and necrophiliac practices of mankind in proclaiming this common blood-lust differentiator, this predaceous habit, this mark of Cain that separates man dietetically from his anthropoidal relatives and allies him rather with the deadliest of Carnivora.

Robert Ardrey was an American film script writer who, having run afoul of the McCarthy anti-communist hearings, was so taken by Dart's speculations that he moved to Africa. He put his talents to good use for he became a best-selling writer about human origins, pushing a view of humankind—*African Genesis*—taken directly from Dart. Ardrey was a good writer, serious and respected, but he was an amateur. He was not a serious and respected scientist, one of the fathers of ethology, on the way to a Nobel Prize. In his best-selling *On Aggression*, just such a person, Konrad Lorenz, likewise embraced the killer-ape theme with enthusiasm. We learn that all animals have built-in mechanisms to save them from harming members of their own species. "This is certainly the case in the dog, in which I have repeatedly seen that when the loser in a fight suddenly adopted the submissive attitude, and presented his unprotected neck, the winner performed the movement of shaking to death, in the air, close to the neck of the morally vanquished dog, but with closed mouth, that is, without biting" (Lorenz, 1966, p. 133). With regret, Lorenz has to tell us that humans are the exception. "In human evolution, no inhibitory mechanisms preventing sudden manslaughter were necessary, because quick killing was impossible anyhow; the potential victim had plenty of opportunity to elicit the pity of the aggressor by submissive gestures and appeasing attitudes" (Lorenz, 1966, p. 241). Natural selection let us down. Or rather, natural selection did not prepare us for what was to come. We did not develop an inward distaste against killing each other. Too bad. The invention of artificial weapons was just around the corner. We were helpless: "man's position was very nearly that of a dove which, by some unnatural trick of nature, has suddenly acquired the beak of a raven."

That seems to be that then. Except it isn't. Truly, one ought to be very suspicious of a claim that fits so very nicely with the mood of the age. It may be true, but it is very much the way we would rather like things to be. It's not our fault. Evolution made us into killer apes. What else were we expected to do but go to war and try desperately—and all too successfully—to kill each other? One ought also be suspicious of a claim that meshes so conveniently with Calvinist theology. Dart (1953) gives the game away. He starts his piece with a theological reflection. "Of all beasts the man-beast is the worst, To others and himself the cruellest foe." Quoting *Christian Ethics* by Richard Baxter (1615–1691). Compare Zechariah 8:10 "For before those days there was no wage for man or any wage for beast, neither was there any safety from the foe for him who went out or came in, for I set every man against his neighbor." Remember also, "this common bloodlust differentiator, this predaceous habit, this mark of Cain that separates man dietetically from his anthropoidal relatives." This is not science. We are getting the wisdom of the bible rather than the wisdom of the rocks. We are on the way to the Augustinian thesis of original sin. We are all tainted because of our first ancestors. Eve, then Adam, disobeyed God and ate that wretched apple. Original sin! Killer apes!

There is much to make one wary of Dart's gloomy speculations. Most importantly of all, it seems that the killer ape hypothesis is empirically false! Accounts of actual warfare strongly suggest that the urge and ability to kill our fellow humans has little solid evidence in its favor. For most of us, such an urge is nonexistent! "During World War II U.S. Army Brigadier General S. L. A. Marshall asked these average soldiers [whom he was interviewing] what it was they did in battle" (Grossman, 2009, p. 3). His findings were staggering: "Of every hundred men along the line of fire during the period of an encounter, an average of only 15 to 20 'would take any part with their weapons.' This was consistently true 'whether the action was spread over a day, or two days or three.'" Marshall stressed that the non-firing soldiers were not wimps or cowards. To the contrary. Queries to those who had recently been in close combat with either European or Japanese foes: "The results were consistently the same: only 15 to 20 percent of the American riflemen in combat during World War II would fire at the enemy." However, "Those who would not fire did not run or hide (in many cases they were willing to risk great danger to rescue colleagues, get ammunition, or run messages), but they simply would not fire their weapons at the enemy, even when they were faced with repeated waves of banzai charges" (pp. 3–4).

Unsurprisingly, measured in terms of psychological stress, medics are in very good shape. Their role in war is healing people not killing them. A paradox. Officers emerge in a pretty good mental state. You might think that they, above others, would be stressed out. Not so. "Indeed, it is a generally accepted tenet of modern warfare that if an officer is shooting at the enemy, he is not doing his job." Marshall stressed that this is not a function of cowardice. "He led his soldiers from behind, he found it less exciting." In the First World War, 27% of officers serving on the Western Front were killed, as opposed to 12% of the men. It is rather that he doesn't have the psychological strain of killing another human being. "Over the men, officers in that war had a 50% less chance of being psychologically disabled" (p. 63).

So how does killing occur? Most obviously and commonly at a distance. In the Vietnam War, Lieutenant William Calley leads his troops into the village of My Lai, where they killed 500 unarmed people, mainly women and children, having first gang raped some of the women and then mutilated their genitalia in disgusting ways. The horror and shock at such behavior still resonates 50 years later. Normal human beings, normal soldiers even in wartime, simply don't do that sort of thing. However, equally normal people put in a Lancaster bomber flying over Germany have no hesitation in dropping bombs on civilian populations below, killing at least as many people as did Calley and his thugs. It is the same at sea. People manning the guns on destroyers have no hesitation in shooting at the enemy. The death and destruction comes at a distance. Essentially, humans can only do this sort of thing when they are separated from their actions. Again, the stress is on the fact that this has nothing to do with cowardice. The very bravest of men often refuse or are constitutionally unable to shoot and kill a visible enemy. At the individual level, we are just not mad murderers. And this speaks for the best of us at least, if not more, than for the others.

Prejudice

Let's pull back and start again. Humans are not killer apes. Indeed, if we are not being dramatic, or sounding like Scottish Presbyterians on a roll, there is masses of evidence to show otherwise. There is, however, the other side to us. Let us not sentimentalize. The Great War, the First World War (1914–1918), ended with twenty to forty million dead; the Second World War (1939–1945) – sixty to eighty million dead; the Russian Civil War (1917–1922) – five to ten million dead; and the Chinese Civil War (1927–1949) – about ten million dead. And this all before you get to genocide. During the Great War and for some years after, over three million Armenian Christians were murdered by the Turks. Let us not forget the Kulaks, the wealthy peasant farmers, whom Stalin regarded as enemies of the Soviet state. In the 1930s, at least a million were liquidated, perhaps more. An equal number died because of the resulting famine caused by the elimination of the Kulaks. Hovering over us all is the dreadful tale of the Germans and the Jews—at least six million deaths. The Holocaust will forever be a stain on humankind (Friedlander, 1997, 2008). Europeans are not the only ones. In Rwanda, in 1994, up to a million Tutsis were murdered and up to half a million Tutsi women were raped, often as a preliminary to grotesque mutilations of their genitalia.

War (and genocide) are only one part of the story of humans being so nasty to each other. There is prejudice—attitudes as opposed to actual killing. One need only mention Brexit to bring home how much the English hate foreigners. Not something new I am afraid. Shakespeare (*Richard II*) played on it.

This royal throne of kings, this sceptred isle,
This earth of majesty, this seat of Mars,
This other Eden, demi-paradise,

This fortress built by Nature for herself
Against infection and the hand of war,
This happy breed of men, this little world,
This precious stone set in the silver sea,
Which serves it in the office of a wall
Or as a moat defensive to a house,
Against the envy of less happier lands,--
This blessed plot, this earth, this realm, this England.

There is little surprise that the God of the English speaks in the dialect of the King James Version of the Holy Bible. Or that this attitude lasted right through the centuries to the present. Brexit did not come from nowhere. This is from Charles Dickens's great novel *Little Dorrit* published between 1855 and 1857. He is writing of people in a working-class part of London, and of how it "was uphill work for a foreigner" to make any progress toward social acceptance.

> In the first place, they were vaguely persuaded that every foreigner had a knife about him; in the second, they held it to be a sound constitutional national axiom that he ought to go home to his own country. They never thought of inquiring how many of their own countrymen would be returned upon their hands from divers parts of the world, if the principle were generally recognised; they considered it particularly and peculiarly British. In the third place, they had a notion that it was a sort of Divine visitation upon a foreigner that he was not an Englishman, and that all kinds of calamities happened to his country because it did things that England did not, and did not do things that England did. (Dickens, 1857, p. 302)

Enough said.

Hunter-Gatherers: War

How do we tackle the paradox? Humans loving. Humans hateful. Since Charles Darwin and his *Origin of Species*, 1859, our thread of Ariadne is a given. As Thomas Henry Huxley used to say, "We are modified monkeys, not modified mud." Start with human history (Fry, 2013). Between 5 and 7 million years ago, we parted company with our fellow apes and moved out of the jungle and onto the plains. Instead of foraging for fruits and the like from trees, we became hunter-gatherers. At once, there were new demands on natural selection. We became bipedal, at least in part probably so we could see distances and spot predators/prey. Also being bipedal means you are never going to be that fast—try beating a small dog like a terrier on the flat—but you can outlast the exhausted prey until you can move in for the kill. What does this imply? A 1000 protohumans are never going to function well together. For real efficiency, as we engaged in what has been called our 5-million-year camping trip, we need bands of about 50, whose philosophy of life must be that of Benjamin Franklin after he signed the Declaration of Independence: "Gentlemen, we must all hang together or assuredly we shall all hang separately." We had to be social, both positively—able to get on with each other—and negatively—no claws or fangs or like weapons of attack. "Indeed, given the lack of fearsome teeth and other morphological or technological means to kill one another, the arguable probability is that they lived consistently in groups with sizable cohorts of adult males under an omnipresent possibility that they might be confronted by carnivoran predators, which were more formidable in number and variety than modern ones" (Tuttle, 2014, p. 593).

What about war? Well, what about war? The lands—a first Africa and then broadening out across the globe—were vast. And hominins (human ancestors) were few. "Just prior to the Holocene (19,000 to 13,000 BP), the Late Pleistocene populations of Australia, Asia, Europe, and Africa combined have been estimated at about 500,000 people, or .3 persons per 100 sq. km" (Haas & Piscitelli, 2014). Going to war with another group, tribe, means you might get hurt. Far better just to move away. Plenty of room for everyone. The point is that there was just no selective reason to be killer apes. Far better to focus on your own group and be friendly and responsible.

So, what happened? In a word agriculture! Around 10,000 years ago, humans started into agriculture in a big way. That meant, on the one hand, fixed possessions. You cannot pick up your farm and move away. On the other hand, a huge population explosion. In a hunter-gatherer band of 50, numbers of children must be limited—by infanticide and the like—or the band could not function. Apart from anything else, the females would be taken out of action as they gave birth to and raised offspring. With agriculture, the picture changes: "farmers pump out babies much faster than hunter-gatherers" (Lieberman, 2013, p. 188). How come? "After a few years of care, a farmer's children can work in the fields and in the home, helping to take care of crops, herd animals, mind younger children, and process food. In fact, a large part of the success of farming is that farmers breed their own labor force more effectively than hunter-gatherers which pumps energy back into the system, driving up fertility rates."

The way is opened for war and hatred. "In most formulations, some basic parameters appear: higher populations, more sedentism, foraging concentrated on spatially limited and highly productive sites, food storage, definition of more distinctive social groups, and sociopolitical hierarchy. These all apply, with local variations, to the European Mesolithic" (Ferguson, 2013b, p. 199). Fixed possessions; insufficient room. Is there evidence for any of this? Is there evidence against any of this? Positively, there is the archeological record. There is simply no evidence of war before about 10,000 years ago, the time of the start of agriculture. This does not mean that humans were never violent. There were no automobiles, so no road rage. But there were other people's wives. There were old folk who simply were no longer contributing to their keep. There were psychopaths who were going to be dangerous, whatever the situation. None of this is war. Which simply did not exist. In the words of anthropologist Brian Ferguson, "We are not hard-wired for war. We learn it" (Ferguson, 2013a, p. 126).

> True, in some cases, war could be present but not leave any traces. However, comparison of many, many cases, from all different regions, shows some clear patterns. In the earliest remains, other than occasional cannibalism, there is no evidence of war, and barely evidence of interpersonal violence. In Europe's Mesolithic [15,000–5,000], war is scattered

> and episodic, and in the comparable Epipaleolithic [20,000–8,000] in the Near East it is absent. (Ferguson, 2013b, p. 191)

What about counter-evidence? Looking at today's hunter-gatherers, it seems that the Yanomamö in Brazil give the complete lie to this rosy picture.

> Studies of the Yanomamö Indians of Amazonas during the past 23 years show that 44 percent of males estimated to be 25 or older have participated in the killing of someone, that approximately 30 percent of adult male deaths are due to violence, and that nearly 70 percent of all adults over an estimated 40 years of age have lost a close genetic relative due to violence. Demographic data indicate that men who have killed have more wives and offspring than men who have not killed. (Chagnon, 1988, p. 985)

Claims like these by anthropologist Napoleon Chagnon are, as they say, "hotly contested." critics. First, Chagnon is not talking about what we have been talking about, small hunter-gatherer groups with lots of space: "pre-Columbian native societies of northern Amazonia were not characterized by the small size and simple organization of ethnographically known shifting horticulturalists." In fact, "Ethnohistory and archaeology reveal large riverine settlements, bordering on urban scale, linked together with connections of trade, marriage, war, alliance, and ritual. These systems reached into smaller scale societies in highland interiors. Without any doubt, these were social worlds full of tumult, change, and conflict" (Ferguson, 2015, p. 382). Second, there has been much in the past few centuries to make the Yanomamö totally irrelevant. "Slave raiding by or for Europeans reached the Negro, Branco, and Orinoco rivers that encompass Yanomami highlands from the 1620s onward, more or less rapidly devastating lowland societies." Adding, "In the 1730s, captive-taking reached a peak, with some 20,000 going to the Portuguese from 1740 to 1750. These wars for captives totally transformed the indigenous world of northern Amazonia" (p. 382). In short, the Yanomamö "do not represent humankind in a state of nature, but were mightily 'touched' by the imperial world for centuries. The society Chagnon studied was drastically restructured by western contact, both within and between local groups."

If not today's hunter-gatherer groups, what about the apes? Thanks to long-term studies of chimpanzees in Africa by Jane Goodall (1986), the evidence is strong that chimpanzees are indeed killer apes. "As the number of chimpanzee study sites has increased across Africa, and with accumulating information from long-term study sites, it has become clear that intraspecific aggression constitutes a pervasive risk for chimpanzees" (Wilson, 2013, p. 370). Some killing occurs within groups. However, "the majority of killings (67 percent) involved intergroup attacks… Intergroup killing thus appears to be a widespread trait of chimpanzees, rather than the result of circumstances peculiar to one or a few study sites" (Wilson, 2013, p. 370).

Yes, but for a start, hominins are not chimpanzees. They were out on the plains with masses of room, not stuck in jungles where there might indeed be conflict for resources. Obviously, physically, we have evolved away from chimpanzees. Why should not the same be true of our social nature? For a second, there are the bonobos—pygmy chimpanzees. Their rule is that of the hippies in the 1960s. Make love not war. "Male bonobos do not form the tight bands that are associated with the male cooperative killing behavior of chimpanzees. Instead, bonobo aggression is mild. Disputes and social tensions among bonobos are often diffused through sexual behavior" (White et al., 2013, p. 392). There are various speculations as to why chimpanzees and bonobos are so different behaviorally. Suggestive is the fact that bonobos live in areas where there would not be fierce competition for foodstuffs. One study found that whereas 37 percent of the diet of bonobos was of low-lying plants ("terrestrial herbaceous vegetation"), relatively easy to access, only 7% of the chimpanzee diet was of comparable stuffs. Also probably relevant is the fact that bonobos (like hominins) live in bigger areas than do humans. Getting out of the way is a real option.

Hunter-Gatherers: Prejudice

Thus, the paleoanthropological explanation for war. It is not innate. We learn it, something which started with the development of agriculture 10 millennia ago. What about prejudice? Expectedly, it is connected. For fairly obvious reasons, the big need of hunter-gatherers is to get on with their fellows (ingroup). Dislike of outsiders is secondary (outgroup).

> Since ingroup/outgroup distinctions do not always involve intense (or even mild) competition or conflict over scarce resources, there is need for a theory of the evolution of social groups that does not depend on intergroup conflict per se. Such a theory starts from the recognition that group living represents the fundamental survival strategy that characterizes the human species. In the course of our evolutionary history, humans abandoned most of the physical characteristics and instincts that make possible survival and reproduction as isolated individuals or pairs of individuals, in favor of other advantages that require cooperative interdependence with others in order to survive in a broad range of physical environments. In other words, as a species we have evolved to rely on cooperation rather than strength, and on social learning rather than instinct as basic adaptations. The result is that, as a species, human beings are characterized by obligatory interdependence. (Brewer, 1999, p. 433)

Notice that this says nothing about outgroup as such. The first and primary need is ingroup cohesion.

> The evolutionary argument for bounded social cooperation carries no implicit link between ingroup formation and intergroup hostility or conflict. In fact, in a context of limited resources, group differentiation and territory boundaries can serve as a mechanism to prevent conflict among individuals rather than promoting it. Discrimination between ingroup and outgroups is a matter of relative favoritism toward the ingroup and the absence of equivalent favoritism toward outgroups. Within this framework, outgroups can be viewed with indifference, sympathy, even admiration, as long as intergroup distinctiveness is maintained. (Brewer, 1999, p. 434)

Backing this kind of thinking, hunter-gatherer groups in Australia, faced with very difficult climate and circumstances, are good at getting along together and helping others. It is what is known as "reciprocal altruism." You scratch my back and I will scratch yours. "In Australia's Western Desert, it would seem that a harsh environment with unpredictable rainfall creates mutual dependence among different groups." As a consequence, "This interdependence, in turn, facilitates peaceful intergroup relations that encourage the desert-dwellers to cooperate, offer assistance to each other in times of need, reciprocally share resources, and behave kindly toward others within and beyond the immediate group" (Fry, 2014, p. 87).

However, it can be seen all too easily how prejudice such as fueled Brexit can arise. Britain, a fertile island, is all too tempting a prize. There have been repeated invasions in the past—Julius Caesar (55–54 BCE) and William the Conqueror (William of Normandy) (1066 CE) most famously. Earlier invasions included the Beaker Folk, so named because of the style of their pottery. They arrived around 4000 years ago, kicking out the residents. "The genetic impact of the spread of peoples from the continent into the British Isles in this period was permanent. British and Irish skeletons from the Bronze Age that followed the Beaker period had at most about 10 percent ancestry from the first farmers of these islands, with the other 90 percent from people like those closely associated with the Bell Beaker culture in the Netherlands" (Reich, 2018, p. 115). Today's druids and the like may celebrate the Summer Solstice at Stonehenge. It was not their ancestors who built it.

Rudyard Kipling put it well.

The Stranger within my gate,
He may be true or kind,
But he does not talk my talk--
I cannot feel his mind.
I see the face and the eyes and the mouth,
But not the soul behind.

The men of my own stock,
They may do ill or well,
But they tell the lies I am wanted to,
They are used to the lies I tell;
And we do not need interpreters
When we go to buy or sell.

The Stranger within my gates,
He may be evil or good,
But I cannot tell what powers control--
What reasons sway his mood;
Nor when the Gods of his far-off land
Shall repossess his blood.

The men of my own stock,
Bitter bad they may be,
But, at least, they hear the things I hear,
And see the things I see;
And whatever I think of them and their likes
They think of the likes of me.

This was my father's belief
And this is also mine:
Let the corn be all one sheaf--

And the grapes be all one vine,
Ere our children's teeth are set on edge
By bitter bread and wine.

One can, of course, see how prejudice could arise within ingroups. Take disability. A priori one might think that the disabled are going to bring on feelings of pity and a desire to help. In fact, often the very opposite is the case. We are disgusted by disability and reject the sufferers. Another poet, this by Wilfred Owen, one of the greatest of the (Great War) poets.

He sat in a wheeled chair, waiting for dark,
And shivered in his ghastly suit of grey,
Legless, sewn short at elbow. Through the park
Voices of boys rang saddening like a hymn,
Voices of play and pleasure after day,
Till gathering sleep had mothered them from him.

About this time Town used to swing so gay
When glow-lamps budded in the light-blue trees
And girls glanced lovelier as the air grew dim,
—In the old times, before he threw away his knees.
Now he will never feel again how slim
Girls' waists are, or how warm their subtle hands,
All of them touch him like some queer disease.

There was an artist silly for his face,
For it was younger than his youth, last year.
Now he is old; his back will never brace;
He's lost his colour very far from here,
Poured it down shell-holes till the veins ran dry,
And half his lifetime lapsed in the hot race,
And leap of purple spurted from his thigh.
One time he liked a bloodsmear down his leg,
After the matches carried shoulder-high.
It was after football, when he'd drunk a peg,
He thought he'd better join. He wonders why …
Someone had said he'd look a god in kilts.

That's why; and maybe, too, to please his Meg,
Aye, that was it, to please the giddy jilts,
He asked to join. He didn't have to beg;
Smiling they wrote his lie; aged nineteen years.
Germans he scarcely thought of; and no fears
Of Fear came yet. He thought of jewelled hilts
For daggers in plaid socks; of smart salutes;
And care of arms; and leave; and pay arrears;
Esprit de corps; and hints for young recruits.
And soon, he was drafted out with drums and cheers.

Some cheered him home, but not as crowds cheer Goal.
Only a solemn man who brought him fruits
Thanked him; and then inquired about his soul.
Now, he will spend a few sick years in Institutes,

And do what things the rules consider wise,
And take whatever pity they may dole.
To-night he noticed how the women's eyes
Passed from him to the strong men that were whole.
How cold and late it is! Why don't they come
And put him into bed? Why don't they come?

Heartbreaking. And much a function of rejection. "To-night he noticed how the women's eyes. Passed from him to the strong men that were whole." Natural selection is not your friend. In a small band, dependent on everyone pulling their weight, a man like this would simply be a burden. He would put the whole group in danger. Hence, we must reject him and, to make sure we reject him, selection has picked out those of our would-be ancestors who did feel a sense of disgust—although, fear is often a major factor. Note also how often we justify our emotions and actions by ascribing ill intent to the disabled.

Now is the winter of our discontent
Made glorious summer by this sun of York;

Richard (to be crowned Richard the Third) about his brother Edward (Edward the Fourth). Richard, however, has other ideas and first admits to his deformity.

I, that am rudely stamp'd, and want love's majesty
To strut before a wanton ambling nymph;
I, that am curtail'd of this fair proportion,
Cheated of feature by dissembling nature,
Deformed, unfinish'd, sent before my time
Into this breathing world, scarce half made up,

Then he admits to his evil nature and how he intends to make use of it.

I am determined to prove a villain
And hate the idle pleasures of these days.

And he certainly does, starting with killing the little princes in the tower.

There are many other instances and types of prejudice—class, race, sexual orientation, religion, and more. Let us take one more example, that of females. All the evidence we have of hunter-gatherers is that women needed to pull their weight—and they did. No weak sisters, to give this rather derogatory term a literal reading. As has been explained, they tended not to be overburdened with children—that came with agriculture. There is absolutely no reason why they should not have been as intelligent as men. While the men were out hunting, pursuing large game with weapons like spears, women were building traps to ensnare smaller prey. And doing their share of gathering and hut building and all the rest that was needed. Conversely, men had to do their share.

> A possible clue for the evolution of sex equality in the hominin lineage was the increase in the cost of human reproduction associated with larger brain sizes in early Homo. Higher offspring costs would require investment from both mothers and fathers, as seen among extant hunter-gatherers. The need for biparental investment predicts increased sex equality, which is reflected in the high frequency of monogamy and the reproductive schedules of male hunter-gatherers who typically stop reproducing early and exhibit long life spans after their last reproduction. (Dyble et al., 2015, p. 798)

Then came agriculture and babies and babies and babies, and it was all over for the status of women. Men assumed power and they used it.

Reversal? War

What do we make of all of this? The main message surely is that often—too often—we are not very nice people. However, the root of the not-niceness does not lie directly in our genes. There is no gene for being warlike as there is a gene for having blue eyes. It had to be triggered, as it was with the coming of agriculture. Hence, there is hope of reversal, without having to get into plans for wholesale genetic engineering. Start with war. In the past 2000 years, there has been much work by philosophers and others—theologians, political scientists, and the like—on spelling out conditions for war, trying to see if, and when, it is justified and if, and when, it is not justified. "Just War Theory," starting with Cicero (106–43 BCE), going through Augustine (354–430 CE) and Aquinas (1225–1274 CE), and then on to Hugo Grotius (1583–1645 CE) and others.

Setting the scene, Cicero slides easily back and forth between *jus ad bellum* and *jus in bello*. Going to war without provocation falls under *jus ad bellum*. But then, conduct in the war, *jus in bello*, is equally important. "Not only must we show consideration for those whom we have conquered by force of arms but we must also ensure protection to those *who* lay down their arms and throw themselves upon the mercy of our generals, even though the battering-ram has hammered at their walls" (*De Officiis*, in Holmes, 2005, p. 29). People started to elaborate and refine. Augustine on *jus ad bellum*. "Peace should be the object of your desire; war should be waged only as a necessity, and waged only that by it God may deliver men from the necessity and preserve them in peace. For peace is not served in order to the kindling of war, but war is waged in order that peace may be obtained" (*Letter to Boniface*, in Holmes, 2005, pp. 62–63). Then on to *jus in bello*, "Let necessity, therefore, and not your will, slay the enemy who fights against you. As violence is used against him who rebels and resists, so mercy is due to the vanquished or the captive, especially in the case in which future troubling of the peace is not to be feared" (p. 63). A refinement from Aquinas, "First, the *authority* of the sovereign by whose command the war is to be waged. For it is not the business of a private individual to declare war, because he can seek for redress of his rights from the tribunal of his superior" (p. 107). So to Grotius, "If not for Justice, yet for Pity, we must not attempt any Thing which may prove the Destruction of Innocents, unless for some extraordinary Reasons, and for the Safety of many" (Grotius, 1625, p. 3, VIII). Hence, "Tender Age must excuse the Child, and her Sex the Woman" (p. 3, IX, 1). We must spare "those who apply themselves to the Study of Sciences and Arts beneficial to Mankind." Philosopher professors heave a great sigh of relief! And more refinements. "Honour, that is, a regard to Equity, does not permit us to take away the Life of a Prisoner" (p. 3, XIII, 1). Grotius adds, "The same Equity commands us to spare those, who surrender to the Conqueror without Conditions in a suppliant Manner. To kill those that have yielded, (says Tacitus) is barbarous" (p. 3, XV).

Do we see any application of these rules and conditions? Take the two World Wars. One can certainly say of the First that it did not fulfill any Just War conditions for its start (MacMillan, 2014). The Archduke of Austria was assassinated in Sarajevo, Austria issued an ultimatum to Serbia, Russia mobilized, and Germany jumped in. Others followed. Britain entered because Germany invaded Belgium. "Peace should be the object of your desire; war should be waged only as a necessity, and waged only that by it God may deliver men from the necessity and preserve them in peace." Not much of that from Austria, Russia, or Germany. Britain perhaps. They did not do it from love of the Belgians, whom they loathed and detested because of their treatment of natives in the African Congo. Then the waging of war. No nonsense in Britain about a Just War. Britain was fighting a "Holy War." In the case of a just war, there are disinterested moral rules to be obeyed. For the former, anything goes so long as it is in the interests of the deity or religious position being endorsed.

> To save the freedom of the world, to save Liberty's own self, to save the honour of women and the innocence of children, everything that is noblest in Europe, everyone that loves freedom and honour, everyone that puts principle above ease, and life itself beyond mere living, are banded in a great crusade — we cannot deny it — to kill Germans. (Marrin, 1974, p. 175)

Thus, Arthur Winnington-Ingram, Bishop of London, "Bishop of the Battlefields." Not that the Germans were any better. "German Christianity represents the right relation between Christ and His disciples, and our nature the most perfect consummation of Christianity as a whole. We fight, then, not only for our land and our people; no, for humanity in its most mature form of development; in a word, for Christianity as against degeneration and barbarism" (Bang, 1917, p. 69).

The treatment of civilians was appalling. In Belgium, in came the Germans. "The small industrial town of Tamines on the Meuse saw 383 inhabitants killed on 22 August; the city of Dinant, on 23 August, where the worst massacre of the invasion left 674 people, one out of every 10 inhabitants, dead; and the university town of Louvain (Leuven), where the treasured university library was burned and 248 civilians killed" (de Schaepdrijver, 2014). In the same mode, God helped soldiers who tried to surrender. And the methods. Tear gas, chlorine, phosgene, mustard gas. "It is a cowardly form of warfare which does not commend itself to me or other English soldiers." Nevertheless, since winning "can only be done by our copying the enemy in his choice of weapons, we must not refuse to do so" (Cook, 1999, p. 37).

The Second World War likewise saw flagrant disregard for Just War Theory. On the Allied side, obliteration bombing. The American Jesuit, John C. Ford, drew attention to its effects: "The total weight of the bombs dropped on Hamburg in seven days equaled the tonnage dropped on London during the whole of the 1940-1941 blitz" (Ford, 1944, p. 293). Continuing, an RAF commentator said, "To all intents and purposes a city of 1,800,000 inhabitants lies in absolute ruins.... It is probably the most complete blotting-out of a city that ever happened." He hardly had to argue that this was a gross—grotesque—violation of Just War Theory. Remember Grotius, "If not for Justice, yet for Pity, we must not attempt any Thing which may prove the Destruction of Innocents, unless for some extraordinary

Reasons, and for the Safety of many" (p. 3, VIII). Even then, people knew that such bombing never had the desired effects. The British were even stronger against Hitler after the Blitz of 1940–1941 than before. Enough information was coming out of Germany to know that industrial production was going *up*, rather than declining. As Ford concluded, "It is fundamental in the Catholic view that to take the life of an innocent person is always intrinsically wrong, that is, forbidden absolutely by natural law. Neither the state nor any private individual can thus dispose of the lives of the innocent" (p. 272).

On both sides, there were other egregious violations of Just War Theory. The Germans, in the early parts of the War, against civilians in Poland and further East. Then, toward the end of the War, the Russians as they moved into East Prussia—unbelievable behavior. Non-stop rape of the women for a start. Non-stop rape encouraged by their superiors. This all said, there were significant improvements from the First World War to the Second. As noted already, it is hard to see real justification for most of the combatants joining the First World War. In the Second, Hitler invaded Poland—after significant efforts to prevent war, the Munich Agreement of 1938 for example. For the Allies, if anything was a clear example of a justified defensive war, it was this. Even the demand for unconditional surrender was fully justified. All efforts had to be made to avoid the unsatisfactory ending of the First World War, not to mention total revulsion at the grotesquely immoral Nazi state. "A just war is wont to be described as one that avenges wrongs, when a nation or state has to be punished, for refusing to make amends for the wrongs inflicted by its subjects, or to restore what it has seized unjustly," Aquinas quoting Augustine.

Treatment of prisoners and of those surrendering also. Not universally. The Nazis behaved appallingly to those whom they conquered. In the First World War, the British had much for which to answer.

> A brigadier was heard by a soldier in the Suffolks to say on the eve of the battle of the Somme: 'You may take prisoners, but I don't want to see them.' Another soldier, in the 17th Highland Light Infantry, recalled the order 'that no quarter was to be shown to the enemy and no prisoners taken'. Private Arthur Hubbard of the London Scottish also received strict orders not to take prisoners, 'no matter if wounded'. His 'first job', he recalled, 'was when I had finished cutting some of the wire away, to empty my magazine on 3 Germans that came out of their deep dugouts, bleeding badly, and put them out of their misery, they cried for mercy, but I had my orders, they had no feelings whatever for us poor chaps'. (Ferguson, 2004, p. 158)

In the Second World War, things did start to change. Take the case of Major John Cochran, in March 1945, faced with a 16-year-old German boy, who, before surrendering, had killed one of his men. "I was very emotional over the loss of a good soldier and I grabbed the kid and took off my cartridge belt. I asked him if there were any more like him in the town. He gave me a stare and said, 'I'd rather die than tell you anything.'" Cochran started to beat him up, "when I was grabbed behind by Chaplain Kerns. He said, 'Don't!' Then he took the crying child away. The chaplain had intervened not only to save a life but to prevent me from committing a murder. Had it not been for the chaplain, I would have" (Ambrose, 2002, p. 548). "Prevent me from committing a murder." Cicero would have approved.

Reversal? Prejudice

Turn now to prejudice. It is possible for things to change for the better and there is reason for, if not optimism, a modicum of justified hope? Brexit shows only too well—and the election of Trump with his call for excluding walls gives strong confirmation—that prejudice against foreigners is alive and well. Although, perhaps, "well" is not quite the term we need here. What about other areas? There are two keys for change. Social conditions—improved social conditions. Knowledge—new knowledge. The Germans did not have genes that made them uniquely persecutors of the Jews. It was because they were genetically open to prejudice. This was combined with huge societal pressures brought on by resentment at the Versailles Treaty and then the Depression. And by the idea, going back at least to Martin Luther, of the Jews as a threat to decent German life. "The Final Solution did not develop as evil incarnate but rather as the dark side of ethnic righteousness" (Koonz, 2003, p. 273).

The point being made about prejudice is illustrated with great clarity by the topic of sexual orientation. In the late nineteenth century, Oscar Wilde, author of *The Importance of being Earnest*, the greatest farce in the English language, was sentenced to 2 years in jail for homosexual activity. Shortly after release, he died in exile in Paris. On February 14, 1995, in that holy of holies—Poets' Corner in Westminster Abbey—Dame Judi Dench read from the play, at the unveiling of a stained-glass window in Wilde's memory. How comes the change? Generally, it is less felt that the function of society is to enforce private morality. Also, we now know that homosexual orientation is not a function of choice in the direction of evil, but a natural, unchangeable disposition of a significant minority of human beings. You cannot change gays (homosexuals), but neither should you fear for every young person who crosses their path. We can now relax and honor the memory of Oscar Fingal O'Flahertie Wilde. (I was born in England in 1940. I doubt there is any educated child of my generation who has not performed in a school production of *The Importance of Being Earnest*—with the English teacher, in drag, playing the dowager Lady Bracknell.)

In the case of sexual orientation, it is knowledge that is the key. Sometimes the key is improved social circumstances. Today, societies can support and help the handicapped. And, overall, they do. For example, the *Americans with Disabilities Act* of 1990 lays out bluntly the needs and challenges, working toward solutions. In the preamble, the Act states as a matter of fact, "physical or mental disabilities in no way diminish a person's right to fully participate in all aspects of society, yet many people with physical or mental disabilities have been precluded from doing so because of discrimination; others who have a record of a disability or are regarded as having a disability also have been subjected to discrimination." A call to action. Not always heeded or carried out to the full. But so very much better than in the past.

Finally, take up the question of women. Equals when hunter-gatherers. Second-class citizens when we turned to agriculture. And now? On the one hand, societal changes make for the emancipation of women. Labor-saving devices freed women from backbreaking duties. There are good reasons to worship at the shrine to the Bendix Washing Machine. Efficient contraception gave them control over their reproduction. No more 10-children families. On the other hand, we simply know now that women are as bright as, if not brighter, than men. Why not? Sitting at home making traps demands at least as much brain power as galivanting across desert plains, clad only in an athletic support, spear in hand.

> Roughly half the [American] population is female, and by most measures they are faring well academically. Consider that by age 25, over one-third of women have completed college (versus 29% of males); women outperform men in nearly all high school and college courses, including mathematics; women now comprise 48% of all college math majors; and women enter graduate and professional schools in numbers equal to most, but not all fields (currently women comprise 50% of MDs, 75% of veterinary medicine doctorates, 48% of life science PhDs, and 68% of psychology PhDs). (Ceci & Williams, 2009, p. 5)

Once again, enough said.

Envoi

It is rare in philosophy papers to end feeling that one has presented a real solution to a real problem. Ask anyone who has worked on the philosophy of mind! This chapter is an exception. Humans are social—think teenagers! And yet, we can be so hateful. War and prejudice. We now know why. Agriculture. And we now know the solutions. They do not lie in schemes for genetic rewiring. Rather, changing circumstances and gaining new knowledge. These are the keys. The doors are not easy to open. But it is now possible.

References

Ambrose, S. E. (2002). The last barrier. In R. Cowley (Ed.), *No end save victory: New Second World War writing*. Putnam Adult.

Ardrey, R. (1961). *African genesis: A personal investigation into the animal origins and nature of man*. Atheneum.

Bang, J. P. (1917). *Hurrah and hallelujah: The teaching of Germany's poets, prophets, professors and preachers*. George H. Doran.

Brewer, M. B. (1999). The psychology of prejudice: Ingroup love or outgroup hate? *Journal of Social Issues, 55*, 429–444.

Ceci, S. J., & Williams, W. M. (2009). *The mathematics of sex: How biology and society conspire to limit talented women and girls*. Oxford University Press.

Chagnon, N. (1988). Life histories, blood revenge, and warfare in a tribal population. *Science, 239*, 985–992.

Cook, T. (1999). *No place to run: The Canadian corps and gas warfare in the First World War*. UBC.

Dart, R. (1953). The predatory transition from ape to man. *International Anthropological and Linguistic Review, 1*(4), 201–217.

Darwin, C. (1859). *On the origin of species by means of natural selection, or the preservation of favoured races in the struggle for life*. John Murray.

de Schaepdrijver, S. (2014). *The German atrocities of 1914*. British Library: World War One. https://www.bl.uk/world-war-one/articles/civilian-atrocities-german-1914

Dickens, C. ([1857] 1948). *Little Dorrit*. Oxford University Press.

Dyble, M., Salali, G. D., Chaudhary, N., Page, A., Smith, D., Thompson, J., Vinicius, L., Mace, R., & Migliano, A. B. (2015). Sex equality can explain the unique social structure of hunter-gatherer bands. *Science, 348*(6236), 796–798.

Ferguson, N. (2004). Prisoner taking and prisoner killing in the age of total war: Towards a political economy of military defeat. *War in History, 11*(2), 148–192.

Ferguson, R. B. (2013a). Pinker's list: Exaggerating prehistory war mortality. In D. P. Fry (Ed.), *War, peace, human nature: The convergence of evolutionary and cultural views* (pp. 112–131). Oxford University Press.

Ferguson, R. B. (2013b). The prehistory of war and peace in Europe and the near east. In D. P. Fry (Ed.), *War, peace, and human nature: The convergence of evolutionary and cultural views* (pp. 191–240). Oxford University Press.

Ferguson, R. B. (2015). History, explanation, and war among the Yanomami: A response to Chagnon's *Noble Savages*. *Anthropological Theory, 15*, 377–406.

Ford, J. C. (1944). The morality of obliteration bombing. *Theological Studies, 5*, 261–309.

Friedlander, S. (1997). *Nazi germany and the jews: The years of persecution 1933–39*. London: Weidenfeld and Nicolson.

Friedlander, S. (2008). *Nazi Germany and the Jews: 1939–1945*. The Years of Extermination. New York: Harper Pereninal.

Fry, D. P. (2013). War, peace, and human nature: The challenge of achieving scientific objectivity. In D. P. Fry (Ed.), *War, peace, and human nature: The convergence of evolutionary and cultural views* (pp. 1–21). Oxford University Press.

Fry, D. P. (2014). Group identity as an obstacle and catalyst of peace. In J. F. Leckman, C. Panter-Brick, & R. Salah (Eds.), *Pathways to peace: The transformative power of children and families* (pp. 79–92). MIT Press.

Goodall, J. (1986). *The chimpanzees of Gombe: Patterns of behavior*. Belknap.

Grossman, D. (2009). *On killing: The psychological cost of learning to kill in war and society*. Back Bay Books.

Grotius, H. (1625 [1901]). The rights of war and peace, including the law of nature and of nations (A. M. Campbell, Ed.). M. Walter Dunne.

Haas, J., & Piscitelli, M. (2014). The prehistory of warfare: Misled by ethnography. In D. P. Fry (Ed.), *War, peace, and human nature: The convergence of evolutionary and cultural views* (pp. 168–190). Oxford University Press.

Hersch, S. M. (1972). *Cover-up: The army's secret investigation of the massacre at My Lai 4*. Random House.

Holmes, A. F. (Ed.). (2005). *War and Christian ethics* (2nd ed.). Grand Rapids, Mich.

Koonz, C. (2003). *The Nazi conscience*. Cambridge, Mass.

Lieberman, D. E. (2013). *The story of the human body: Evolution, health, and disease*. Vintage.

Lorenz, K. (1966). *On aggression*. Methuen.

MacMillan, M. (2014). *The war that ended peace: The road to 1914*. Random House.

Marrin, A. (1974). *The last crusade: The church of England in the First World War*. Duke University Press.

Pinker, S. (2011). *The better angels of our nature: Why violence has declined*. Viking.

Reich, D. (2018). *Who we are and how we got here: Ancient DNA and the new science of the human race*. Pantheon.

Tuttle, R. H. (2014). *Apes and human evolution*. Harvard University Press.

White, F. J., Waller, M. T., & Boose, K. J. (2013). Evolution of primate peace. In D. P. Fry (Ed.), *War, peace, and human nature: The convergence of evolutionary and cultural views* (pp. 389–405). Oxford University Press.

Wilson, M. L. (2013). Chimpanzees, warfare, and the invention of peace. In D. P. Fry (Ed.), *War, peace, and human nature: The convergence of evolutionary and cultural views* (pp. 361–388). Oxford University Press.

Michael Ruse is the retired Lucyle T. Werkmeister Professor of Philosophy and Director of the Program in the History and Philosophy of Science at Florida State University. He is also Professor Emeritus at the University of Guelph, in Ontario, Canada, where he taught from 1965 to 2000. He is the founding editor (1985) of the journal *Biology and Philosophy.* Today, he is co-editor of the Cambridge *Elements* series in the Philosophy of Biology. His 70 books include *The Gaia Hypothesis: Science on a Pagan Planet* (Chicago), *Darwinism as Religion: What Literature Tells Us About Evolution* (Oxford), *On Purpose* (Princeton), (co-authored with Edward Larson) *On Faith and Science* (Yale), and, most recently, *Why We Hate: Understanding the Roots of Human Conflict* (Oxford), *Understanding the Life Sciences:* The *Christianity-Evolution Relationship* (Cambridge), and (co-authored with Michael Reiss) *The New Biology: The Battle Between Mechanism and Organicism* (Harvard).

Neoliberalism, Globalization and Discrimination in the Twenty-First Century: Issues, Challenges and Limitations of Habermas' Reason-Based Theory of Liberal Democracy in Current Times

Deborah C. Poff

Abstract This chapter discusses the limitations of Habermas' theory of communicative rationality and deliberative democracy in addressing issues of discrimination as well as addressing issues related to neoliberalism, irrational anti-intellectual mob behaviour and globalization. The focus of the discussion is on the inability of a reason-based theory as the foundation of democracy to address elements of discrimination which are not based on rational understanding.

Keywords Liberal democracy · Discrimination · Diversity · Bias · Rationality · Neoliberalism · Globalization

Introduction

In this chapter, I address the inability of the current key theory of liberal democratic political philosophy to address the challenges adequately, which are posed by diversity theories concerning the nature of discrimination against various groups of people including different races, ethnicities, languages, geography, gender and gender identity. Also, there will be a short discussion of the global concentration of wealth and issues of poverty and hardship in third-world countries (including issues of child labour and slave labour) and the challenge these conditions pose for a theory of deliberative democracy. In addition, the growing anti-intellectual, anti-expertise movements in democratic nation states will be briefly examined.

To undertake such a project is to situate the discussion in a place which some may think is unfair and inappropriate in dealing with philosophical political theory.

D. C. Poff (✉)
Carleton University, Ottawa, ON, Canada
e-mail: PoffD@brandonu.ca

S. Chakraborty (ed.), *Human Minds and Cultures*,
https://doi.org/10.1007/978-3-031-56448-2_6

It is argued that the hypothetical creation and examination of theories of liberal democracy should not be required to be put to the test of application to real conditions in the real world. As Scheuerman (2017) puts this point:

> For uncharitable critics, Habermas remains a native defender of pristine or ideal speech, where actors somehow miraculously engage in rational discourse in a space free of power. With his efforts focused mostly on formulating a rigorous model of idealized communication, Habermas ignores the harsh realities of political life and overstates its rational traits. His theory, they say, seems suited to a philosophical seminar, not the rough and tumble of politics.

Part of the challenge for those uncharitable critics comes from the fulsome and detailed analysis of Habermas' own account of the limitations of nation states, his recognition of the challenges of globalization including economic and environmental problems, and his identification of the need for transnational and supranational mechanisms to address just such rough and tumble politics that face the globe today.

Habermas also recognizes and discusses issues of minority rights and discriminated-against groups. Having granted this, the fundamental issue to which I will return at various points in the discussion is what to do about what some perceive as the ultimate and fundamental paradox between Habermas' theory of discourse-based deliberative democracy in its fully articulated ideal sense and the massive injustices and serious environmental and global challenges facing millions in the world today.

So, while it may be a legitimate point to chastise critics as uncharitable, such response is met with some incredulity by those attempting to address the poverty and discrimination of groups within, in particular, liberal democracies as well as victims of global capitalism inside but especially outside of such democracies.

The chapter necessarily addresses both aspects of the discussion: namely the nature and identity of discriminated-against groups and the theory of liberal democracy currently identified as one of the prominent and leading discussions of democratic theories by Jurgen Habermas, identified by Scheuerman (2017) among others as "our most impressive contemporary theorist of democracy".

First, it should be noted in this discussion that Habermas' theory of communicative action or his theory of deliberative democracy is fundamentally an optimistic theory of hope. The reason I term it as such will I think become self-evident in this chapter as will its limitations in dealing with current political realities. So much has been written about Habermas generally as well as specifically with respect to this theory that scholarly originality here is unlikely if not impossible. There are two tremendous challenges faced by any scholar examining any aspect of Habermasian thought; first, there is the sheer prolific productivity of Habermas, and second, there is the large volume of Habermas' supporters and thoughtful critics.

What I hope to address, partially, is the intentionality of what is inherently an optimistic theory of hope for human agency as well as the limitations, given particular dynamics since the end of the twentieth century, and more strongly, in the twenty-first century. Also, I address some issues and real limitations of alternatives to reason-based theories of democracy.

It should further be noted that the discussion here will not focus on all aspects of Habermas' theory of communicative action and deliberative democracy in detail but will concentrate specifically on his claims about the nature of language, reason and understanding and its role in the founding of society as well as the nature of lifeworld. It is here that what I call the fundamental paradox in Habermas' theory comes into question. This specifically has to do with the analysis of discrimination that I propose here and its relationship to language, reason and understanding. Some may find the exercise in itself partial and perverse as Habermas' theory should not be artificially bifurcated into a discussion about his theory of language, reason and understanding from his conceptualization of society, system and lifeworld. This criticism fails to recognize the importance of the role of mutual understanding and reason in itself as fundamentally important in any theory of deliberative democracy.

Stating the concentration here does not belie the importance of Habermas' theory of state, system or the 'objective' theory of society and the lifeworld. However, communicative action and lifeworld are the focus of the current discussion. As Baxter (2011) puts this point:

> Habermas...introduces the lifeworld not to experience in general nor even to action in general, but at the background and 'horizon' for specifically communicative action. The concept of the lifeworld, Habermas says, is 'complementary to that of communicative action.'

As Habermas states the same matter:

> In coming to an understanding with one another about their situation, participants in communications stand in a cultural tradition which they use and at the same time renew; in coordinating their actions via intersubjective recognition of criticizable validity claims, they rely on memberships in social groups and at the same time reinforce the integration of the latter; through participating in interaction with content reference persons, growing children internalize the value orientations of their social groups and acquire generalized capabilities for action...Under the functionalist aspect of *reaching understanding* communicative actions serves the transmission and renewal of cultural knowledge; under the aspect of *coordinating action*, it serves social integration and the establishment of group solidarity; under the aspect of *socialization*, it serves the formation of personal identities. (2:208)

There have been numerous critiques, amendments and partial revisionings of Habermas' theory of communicative action in light of allegations that Habermas' theory is inherently sexist and racist (among other isms) both in its portrayal of equity, universality and rationality, the lifeworld and the system. Some of those critiques have focused specifically on Habermas' theory of systems which will not be the fulcrum of the discussion here while granting that it is a significant part of his theory of democracy.

Rather the emphasis will involve a discussion of whether discrimination and minorities can be adequately and fairly addressed in terms of equality and social justice within Habermas' exegesis of communicative rationality and its normal operation in the lifeworld.

Given the centrality of the nature and history of discrimination, I will turn now to a discussion of discrimination and then return to the Habermasian theory of language, reason and understanding at a later point.

The Nature of Discrimination

The treatment of discrimination in this discussion will be pragmatic with stipulative definitions and the identification of membership within this definition. This will not involve a philosophical analysis of the history and nature of discrimination. There have been many fine books and articles on just this topic but what I am concerned with in this current discussion is the stipulative parameters of the socio-political identification of discrimination germane to the present discussion of Habermas' theory.

Discrimination is a current frequently prominent concern in policies and procedures of both democratic nation states and private corporations. Short- and long-term mechanisms to eliminate forms of discrimination are among key policies and procedures of many government and corporate entities in democratic nation states currently. Many of those policies and procedures do not examine discrimination in terms of illustrating its determinative nature but take it as a given that widespread discrimination exists while at length outlining mechanisms to intervene and mitigate against its existence. In this discussion, such assumptions are not taken for granted but analysed and discussed again not in a philosophical sense but in an empirical socio-political sense.

In a recent discussion document from the Committee on Publication Ethics on the topic of discrimination in scholarly publishing (2022), the fundamental nature of discrimination was addressed briefly and succinctly. As noted in that document, there is nothing inherent in the concept of discrimination which is harmful, pejorative or negative. In fact, as every student of aesthetics understands, one meaning of discrimination is defined in terms of discernment and distinction. As the Committee on Publication Ethics (COPE) document notes "Discernment here is attributed to education, experience and expertise in judgment and is frequently encountered in the field of aesthetics, for example, where the discerning expert may comparatively evaluate two interpretations of a particular piece of classical music, performed by two different classically trained musicians and find one version superior to the other".

So, distinction and discernment appear not to be inherently problematic examples of discrimination. What then is problematic? Is it a slippery slope between educated discernment and unfounded discrimination? Or is it completely different? Does knowledgeable discernment fall into a completely different category of adjudication from pejorative or biased discrimination? In what follows, an examination of both distinct forms of discrimination and slippery slope forms of discrimination will be identified. As well, two types of discriminated-against groups will be identified: those who claim substantive equality with dominant groups and demand the elimination of differentiated treatment and those who claim different but equal status. Before getting into that discussion, let's see who we are talking about and which groups are generally identified as discriminated-against groups.

COPE (2022) provides a quite inclusive list of discriminated-against groups. Quoting from that document, the groups are:

- Sex/gender identity/LGBTQ2S+
- Race
- Age
- Ethnicity
- Visible minority
- Immigrant minority
- Ability or differently abled
- Disease (including stigmatizing diseases)
- Socioeconomic status
- Literacy and numeracy (lack of)
- Geography
- Health (and illness)
- Religion
- Language and/or culture
- First Nations
- Indigenous people
- De-colonized people (used to refer to Indigenous people and other original citizens of countries which were colonized during exploration and settlement by European explorers and their rulers)
- Vulnerable populations (particularly when considered as subjects or participants in research. Vulnerabilities can be based on transitory or more permanent status). Recently, there has been some movement not to use the term vulnerable as it is deemed stigmatizing and discriminatory in itself.

Further, COPE notes that the concept of intersectionality is generally included in such identifications which references the fact that membership in one group is not mutually exclusive from membership in others. Thus, for example, an Indigenous person can be discriminated against as Indigenous while at the same time also being discriminated against on the basis of sex, age, language and geographic origin and location.

Having identified the large number of groups who are self and other-identified as discriminated-against, let's return to the discussion of how discrimination manifests itself against whom.

Discrimination Against Groups Who Claim Substantive Equality with Dominant Groups in Society

The arguments from these groups have been significantly represented by women and various racial groups (e.g. Black and Hispanic) among others. Much of the research on discrimination against these groups focuses on disparities in representation within occupations where discriminated-against groups or members of groups have the same educational backgrounds and similar work experience but are proportionally underrepresented in the field and compensated differently for the same work.

It is consequently argued that irrelevant, arbitrary and biased criteria are introduced when evaluating characteristics, attributes, qualifications, capacities and abilities of groups or members of groups when compared to the dominant members outside the group who do the evaluations. Further, part of discrimination, it is claimed, is based on the belief that such characterizations of the dominant group as superior may be deemed inherent or a matter of the sociobiological nature of the dominant group rather than circumstantial limitations imposed by historical or cultural history. In this case, research by those who advocate the dominant groups' position believe that discriminated-against groups are biologically/and or psychologically inferior to the dominant group and dismiss claims of systemic discrimination of otherwise equal persons and groups. Those who believe that these discriminated-against groups are equal, on the other hand, cite research that illustrates equal abilities and unequal treatment. For example, Economist Bourguignon (2015) states the discrimination case as follows:

> In most countries, education differentials between men and women have virtually disappeared and, with a higher participation of women in the labor force, job experience has also equalized. The remaining wage differences are thus a result of either discrimination, in a strict sense, women being paid less for exactly the same work, or remaining differences in non-observed characteristics of salaried employees.

This type of discrimination is most easily addressed by Habermas' theory of deliberative democracy because those discriminated against claim to have evidence and reasons which may be tested in terms of validity concerning their fundamental equality and can also make validity claims and present empirical evidence to try to convince others of the legitimacy of their arguments concerning imposed but not essential barriers to equality. I discuss this further on return to Habermas' theory.

Discrimination Against Groups Who Claim Equal but Different Status

A more complex case comes with the politicization of members or groups of discriminated persons who raise claims of equal but different status and argue that they are deserving of equal status in the political nation state while not conceding any need to alter or assimilate to the traits of the dominant group. These cases involve arguments that some "peoples" have different epistemologies, ontologies, metaphysics and axiologies with theories about the nature of the world that they argue are equally worthy of equal status.

The example I choose here as an exemplar is Indigenous or Aboriginal peoples. While it is important not to assume that all Indigenous groups are the same, these claims are sufficiently uniform to be articulated among other places in the United Nations Declaration on the Rights of Indigenous Peoples. Further, others from dominant cultures recognize such equal but different claims and are incorporating

Indigenous perspectives on science, for example, in their teaching and research. In one such example, Robbins (2018) states that Ecologists:

> [a]round the globe, researchers are turning to what is known as Traditional Ecological Knowledge (TEK) to fill out an understanding of the natural world. TEK is deep knowledge of a place that has been painstakingly discovered by those who have adapted to it over thousands of years. 'People have relied on this detailed knowledge for their survival,' Huntington and a colleague wrote in an article on the subject. 'They have literally staked their lives on its accuracy and repeatability.'
>
> This realm has long been studied by disciplines under headings such as ethno-biology, ethno-ornithology, and biocultural diversity. But it has gotten more attention from mainstream scientists lately because of efforts to better understand the world in the face of climate change and the accelerating loss of biodiversity.

Indigenous groups not only agree that their understanding of the environment is sometimes superior to that of the dominant science, but they do so within a belief system that also believes metaphysically in the existence of Mother Earth and believes that damage to the environment harms "Mother Earth" or the "Creator".

As noted above, the United Nations appears clearly to recognize this different but equal status in The United Nations Declaration on the Rights of Indigenous Peoples (2007). Examples of this recognition included in the declaration are the following:

> Article 5. Indigenous peoples have the right to maintain and strengthen their distinct political, legal, economic, social and cultural institutions, while retaining their right to participate fully, if they so choose, in the political, economic, social and cultural life of the State.
>
> Article 13. Indigenous peoples have the right to revitalize, use, develop and transmit to future generations their histories, languages, oral traditions, philosophies, writing systems and literatures...
>
> Article 31. Indigenous peoples have the right to maintain, control, protect and develop their cultural heritage, traditional knowledge and traditional cultural expressions, as well as the manifestations of their sciences, technologies and cultures, including human and genetic resources, seeds, medicines, knowledge of properties of fauna and flora, oral traditions, literatures, designs, sports and traditional games and performing arts. They also have the right to maintain, control, protect and develop their intellectual property over such capital heritage, traditional knowledge, and traditional cultural expressions.

In this example, discriminated-against groups argue that in spite of belief systems that may be considered at odds with norms of rationality, empirical evidence, general principles of understanding and discourse, their claims and their social, economic and political status should be accepted as equal. This form of different but equal poses particular challenges for Habermas' theory and will be discussed further on return to Habermas' theory.

Discrimination: Slippery Slopes or Distinct Forms of Arbitrary Bias or Discrimination

With discrimination and with respect to human beings and with their abilities, distinctions are drawn between individuals or groups in which the argument is made that some are superior to others and consequently more deserving of differential success, compensation or status.

In this first rendering of discrimination, the irrelevant nature of the discrimination is illustrated through rational argument and empirical evidence. With respect to discrimination against women, traditionally, the bias was entrenched in theories about women's inabilities due to their role in reproduction, biologically, psychologically and socially. From the nineteenth century to current times, primarily feminist scholars wrote critiques and provided empirical evidence intended to prove that women were equally qualified to vote, to attend higher education and to work (and work for equal pay). Again, such critiques can be fairly accommodated according to Habermas' theory with the ability to assess the validity of the arguments and empirical evidence in question.

What makes this type of discrimination unfair and unjust is based on the argument and evidence that women and men's ceteris paribus are essentially the same. Consequently, arguments that claim women are inferior to men are deemed discrimination based on the facts.

With respect to the Indigenous example, the challenge is more difficult, in particular for Habermas. This group claims equality status independent of its approach to causality, nature, reason and faith. Moreover, the position is that the tests for discrimination are in themselves morally unjust. While this case of different but equal is made more strongly in this case than it is with respect to other discriminated-against groups, it is the case that many discriminated-against groups do offer a version of different but equal in describing both the grounds of their discrimination and in terms of what equality would mean for them.

A final complexity here also involves a slippery slope between the claims of "otherness" as articulated by Indigenous groups and equality by, for example, women and other races. In the discussion paper by the Committee on Publication Ethics, empirical research is presented which provides data indicating discrimination as follows:

- Women are fewer editors of scholarly journals than men
- Women are fewer members of editorial boards than men
- Women are asked to peer review less frequently than men
- Authors from countries outside of Europe and the United States are less likely to have their articles published, particularly in prestigious or rank A journals
- Authors who are English as a Second Language scholars have fewer articles accepted.

These examples and many more suggest that the peer adjudication of research by members of groups of different gender, race, ethnicity, geography and language may be judged not just on the quality of their research but on other irrelevant, arbitrary and discriminatory grounds. This is where the slippery slope between expertise and bias may prejudice fair evaluation of performance, and it is part of the grounds of complaint by groups who claim unfair discrimination that biases fair and sound treatment of their work when compared to the dominant group that does the judging.

Having introduced the material on discrimination, let us return to Habermas' theory.

Habermas' Theory of Deliberative Democracy

The exposition of Habermas' theory will be brief in this discussion and only serve to contextualize the analysis. Having said this, minimally, it is critical to understand Habermas' position that society is fundamentally grounded in rational understanding and rational argumentation. Of argumentation he states:

> Arguments are the means by which intersubjective recognition of a proponent's hypothetically raised validity claim can be brought about and opinion thereby transformed into knowledge. (1; 25)

Further, he states that:

> ...the discursive process of reaching understanding, in the form of a cooperative division of labor between proponents and opponents, is normatively regulated in such a way that participants
>
> - thematize a problematic validity claim and,
> - relieved of the pressure of action and experience, in a hypothetical attitude,
> - test with reasons, and only with reasons, whether the claim defended by the proponents rightfully stands or not. (1;25)

In describing his theory of communicative action and how it operates in the life world, Habermas states:

> In coming to an understanding with one another about their situation, participants in communication stand in a cultural tradition which they use and at the same time renew; in coordinating their actions via intersubjective recognition of criticizable validity claims, they rely on memberships in social groups and at the same time reinforce the integration of the latter; through participating in interaction with competent reference persons, growing children internalize the value orientations of their social groups and acquire generalized capabilities for action...Under the functional aspect of reaching understanding communicative action serves the transmission and renewal of cultural knowledge; under the aspect of coordinating action; it serves social integration and the establishment of group solidarity; under the aspect of socialization, it serves the formation of personal identities...(2;208)

It is "...how speaking and acting subjects acquire and use knowledge..."

Susen (2017) summarizes Habermas' theory in one short sentence, "...Habermas' discourse ethics constitutes a systematic attempt to locate the normative grounds of deliberative democracy in the rational foundations of language". Susen's purpose in his article is to illustrate how Habermas' theory of language use is "embedded in the pragmatics of interaction" such that "Habermas' concern with democracy is inseparably linked to his interest in language".

Further, it should be noted that in developing his theory, Habermas wanted to provide a meaningful and rational replacement to what he perceived as the limited conception of social contract theorists of the eighteenth century, such as Locke, who perceive human beings as entering into a minimalist societal agreement to negatively restrict other selfish beings from taking what is ours and sets out a trade and barter system of acquisition and negotiation for the exchange of goods that otherwise would be beyond our reach.

Susen goes on to nicely summarize his interpretation of Habermas in terms of its ten elements: (1) deliberation, (2) reciprocity, (3) self-determination, (4) citizenship, (5) the state, (6) sovereignty, (7) communicative rationality, (8) regulation, (9) will formation and (10) constitutional law.

Of all of these elements, the most fundamental relates to communicative rationality. To quote Susen on this point:

> To be exact, the socio-ontological forces of Verstand (reason), Verstandigung (communication), Verstenhen (understanding), Verstandlichkeit (intelligibility), and Einverstandnis (agreement) play a foundational role in the construction of democracy.

Susen notes, however, that there is also no democracy without all aspects of irrationality, "incomprehensibility", "disagreement", "controversy", etc. that also exists.

And, while I stated earlier that I would not focus on the state and life world, it is clear that Habermas intends always to connect rationality to an understanding of an objective, external reality. He says:

> This concept of *communicative rationality* carries with it connotations based ultimately on the central experience of the unconstrained, unifying, consensus-bringing force of argumentative speech, in which different participants overcome their merely subjective views and, owing to the mutuality of rationally motivated conviction, assure themselves of both the unity of the objective world and the intersubjectivity of their lifeworld. (1;10)

In making this claim, Habermas clearly wishes to distinguish his theory from those philosophies stuck in a language game with no analysis or justification for articulating an agreed-upon external reality.

Communication, Equality, Understanding (of Self and Others), Adjudication and Validity Claims

I noted earlier that Habermas' theory is a hopeful one. What creates Habermas' theory as a theory of hope is the stated assumption that the fundamental necessities for democracy and the democratic engagement of citizens is the universally available rational participation as a basis for society to all participants in the state. This is because of the nature of language as Habermas describes it. Language is reason-based; language allows participants to engage in presenting self-interests with reasons that can either be determined to be valid or invalid. Reason allows us not only to understand our own needs and wants but also the needs and wants of others (reciprocity of understanding). These aspects of reason in language allow for deliberative thought, reciprocal understanding and mutual concern and compromise. We do this type of reasoning in the backdrop of our lived experience as fundamentally equal participants by virtue of our equal facility with language, reason and understanding. Importantly, this basis for society does not involve forced imposition by the power of the state.

The hopeful nature of the theory rests in the mutuality and consensus of the reasoning population. It presumes that all people are rational and consequently all can participate in the exercise of the state.

Limitations of the Theory

Habermas' theory rests on a number of challenging assumptions. These include assumptions of equality of rationality; the belief that rationality leads to understanding of the needs of one's self and the consequent understanding of the needs of others; and the ability of those in the engagement to rationally test and confirm or reject the validity of the other's claims. The theory also assumes equality of representation of participants in a rational dialogue. Such assumptions are ideal and hopeful but contrary to much that we know about the nature and foundation of representation of equals in any actual democracy that does or has existed.

Two further elements will be identified as part of the problem. The first is the impact of the globalization of capitalism and the rise of neoliberalism since the 1970s. The second is the rise in right-wing mob behaviour of racist and sexist anti-intellectual groups within democratic nation states.

The globalization of capitalism and neoliberalism since the 1970s

As many scholars have noted, neoliberalism has led to a diminishment of the status of the nation state globally and an expansion of the commodification of workers such that multinational corporations seek out countries with the cheapest available labour to produce their product. This includes child labour and slave labour particularly in resource extraction economies. Multinational corporations are able to establish operations in countries with the lowest salaries, benefits and protections for workers as a means of increasing profitability.

Neoliberalism and globalization have also had an impact on the public services within democratic nation states. As Giroux (2014) notes with respect to the impact of neoliberalism and university education:

> At the level of higher education, the script is similar with a project designed to defund higher education, impose corporate models of governance, purge the university of critical thinkers, turn faculty into a low-wage army of part-time workers, and allow corporate money and power to increasingly decide course content and determine what faculty get hired. As public values are replaced with corporate values, students become clients, faculty are deskilled and depoliticized, tuition rises and more and more working-class and poor minority students are excluded from the benefits of higher education.

One consequence of neoliberalism and globalization has been the extreme concentration of global wealth generally identified as the 1 percent where 1 percent of the world's population holds 45 percent of its wealth (Financial Times, 2017).

The Mobs

Many democratic states have encountered growing movements within those states of ideological violent groups who represent hatred against different races, faiths, ethnicities and religions. The most dominant example of such groups is currently in the United States where mobs of people attempted to overthrow a democratically elected President and government.

The Issue with Globalization, Neoliberalism and the Mobs

The similarity with all of these three matters is that they do not in any way appeal to Habermas' theory of deliberative democracy. Globalization involves the movement of capital and employment around the world in order to increase the wealth of persons and corporations. There is no identifiable understanding and mutually reciprocal belief systems as Habermas describes these operations. Nor is there any exercise in demonstrating the validity of the practices to participants in the world. Similarly, neoliberalism has been and continues to involve national and international drives to privatize goods and services and minimize the existence of public goods. Finally, the mobs do not operate with any reciprocal and rational understanding of individual and collective wants and needs. If any, these groups are recognizable for their irrational and hate-filled prejudice against others.

All three of these movements are resistant to Habermas' theory.

Throughout the history of philosophy, one strong thread has been predicated on a belief that the truth can lead to a rational and just social order. The purpose of this chapter is not to review that literature, so I will focus only on a couple of examples of rational understanding as the foundation for morality and social justice.

Other Forms of Discrimination and Habermas' Theory

Aside from the issues just discussed, earlier in this chapter, the issue of discrimination was identified. Two types of discrimination were identified in that discussion. The first was discrimination based on racism and sexism where there is clear evidence of discrimination and consequently the possibility of engaging in dialogue, reason, evidence and discussion consistent with Habermas' theory of the purpose of language, equality, rationality and evidence to change and eliminate the discrimination. Although it has taken many years for this type of discrimination to be eliminated, it appears to be diminishing in accordance with Habermas' approach to the foundation of society.

More challenging is the second type of discrimination identified as different but equal. The example used was Indigenous persons. However, some brands of feminists, gender and transgender groups make some of the same types of claims. With respect to Indigenous persons, given the grounds and belief systems of these peoples, the basis of granting equality has to be something other than a reasoned, reciprocal and understandable claim for equality. Many Indigenous communities are extremely poor, with few services including clean water. The basis for granting equality becomes something like equality based on just concern rather than equivalent understandings of equality. Such reason is challenging and certainly not easily accommodated by Habermas' account.

Conclusion

Habermas' theory of deliberative democracy as referenced here is a theory, partly grounded in his understanding of equal rational citizens who can come to general understanding from self-understanding. Further, they can found a rational and caring society based on these attributes of human thinking which were tested with respect to discrimination.

What is clear is that not all social, economic and political thought (e.g. neoliberalism) is based on rationality in Habermas' sense but rather on the selfish greed of a minority of the global population. Further, hate-filled political populism is not driven by rationality but rather irrational bias and prejudice.

It is clear that the best of reasoning that we can hope for is as Habermas describes it, but it is not sufficient in nature to be the candidate of choice in eliminating discrimination.

References

Assembly, U. G. (2007). United Nations declaration on the rights of indigenous peoples. *UN Wash, 12*, 1–18.

Baxter, H. (2011). *Habermas: The discourse theory of law and democracy (jurists: Profiles in legal theory.)*. Stanford Law Books.

Bourguignon, F. (2015). *The globalization of inequality (T. Scott-Ralton, Trans.)*. Princeton University Press.

COPE Council. (2022). *COPE discussion document: Diversity and inclusivity*. https://doi.org/10.24318/RL.qSoVsZ.

Giroux, H. (2014). *Neoliberalism's war on higher education*. Haymarket Books.

Robbins, J. (2018, April 26). *Native knowledge: What ecologists are learning from indigenous people*. Yale Environment 360. Yale University: online journal published by the Yale School of Forestry and Environmental Studies.

Scheuerman, W. (2017, April 12). Habermas and the fate of democracy. *Boston Review*.
Susen, S. (2017). *Jurgen Habermas from the Routledge handbook of language and politics*. Routledge. https://doi.org/10.4324/9781315183718.ch3

Deborah C. Poff has a PhD in Philosophy of Science. During her career she variously served as a Director of a Research Institute, a Dean of Arts and Science, a Vice-President Academic and Provost, and a President and Vice-Chancellor at various Canadian universities. She is an active researcher, teacher, and editor and is the founding Editor-in-Chief of the *Journal of Academic Ethics*, which she edited for over 20 years. She was the Editor of the *Journal of Business Ethics* for over 30 years. She is currently the Editor of the *Journal of Scholarly Publishing*, published by University of Toronto Press. Throughout her career, she has been the editor of five scholarly journals and a book series and has published numerous other articles, monographs, and anthologies. In 1995, she was awarded a lifetime honorary membership by the Canadian Research Institute for the Advancement of Women in the category of "Outstanding Contribution to Feminist Scholarship." Her most recent publication is an anthology entitled *University Corporate Social Responsibility and University Governance*. Her monograph in progress is on *Discrimination and Diversity in Business Ethics and Society*. She is Editor-in-Chief of the *Encyclopedia of Business and Professional Ethics*. In 2016, she was awarded the *Order of Canada*.

Legal Emotions: Why Affect Matters for Law

Antara Haldar

Abstract The legal philosopher HLA Hart theorised that what he referred to as the "internal point of view" was a critical component of law. While this thesis is widely accepted, it has not been rigorously empirically investigated. This chapter, linking the jurisprudential literature on the subject with advances in moral psychology and the behavioural, cognitive and affective sciences, deepens the inquiry into the question of what constitutes "the internal point of view". Going beyond the standard accounts of a cost-benefit approach in law and economics and the abrogation to morality in the natural law tradition that trivialises psychological engagement, it presents a novel hypothesis about the mechanics of an individual agent acquiring it. In particular, while "buy in" to institutional systems, especially the law, has been thought of as a cognitive process, this chapter suggests that it is, perhaps, affective purchase that is critical, especially at an early stage of embedding these structures. It argues that this may be at the root of one of the central problems of law and society: legal transplantation, or the institutional chasm between the Global North and Global South that has defined the post–World War II era leading to the "rule of law" being considered almost an exclusive property of the nations of the advanced industrial West.

Keywords HLA Hart · Internal point of view · Law and psychology · Law and emotions · Affect · Legal transplantation

European Research Council; Preparation of this review was facilitated by funding from the European Union's Horizon 2020 research and innovation programme under the Marie Sklodowska-Curie grant agreement No. 844628—LIMEN

A. Haldar (✉)
University of Cambridge, Cambridge, UK

Harvard University, Cambridge, MA, USA

Stanford University, Stanford, CA, USA
e-mail: ah447@cam.ac.uk

S. Chakraborty (ed.), *Human Minds and Cultures*,
https://doi.org/10.1007/978-3-031-56448-2_7

Introduction

The societal role of law has never been more important, but it is in disarray globally. This is true even in what was once the shining beacon of the rule of law, particularly for Common Law countries: the Anglo-American world. In the United States, the Supreme Court ruling in *Dobbs v. Jackson Women's Health Organization*—that reverses *Roe v. Wade* and radically shrinks women's reproductive rights—has ignited a heated debate about the security of civil liberties in the country (Haldar, 2022c), while a spate of shootings in Buffalo, Texas, Philadelphia and elsewhere has sparked conversation about the reform of gun laws; the legitimacy of the Supreme Court is at an all-time low (Haldar, 2022b) and the former American President, Donald Trump, is implicated increasingly directly in undermining the rule of law in the riots at the Capitol (Haldar, 2022d) and beyond. In the United Kingdom, a messy divorce from Europe in the form of Brexit has been highly disruptive for its legal system, and support for a series of the nation's Prime Ministers is at low ebb, whether as a result of being too libertine (Haldar, 2022a) or libertarian (Haldar, 2022e).

The melee is a wider one. In the international domain, Russia is blatantly undermining international law with its invasion of Ukraine, while the spectre of the existential threat of Climate Change looms large. Meanwhile, the after-effects of the Global Pandemic, the Global Financial Crisis and the Eurozone Crisis linger, and the abiding legacies of hundreds of years of colonialism, slavery and sexism persist, resulting in a highly unequal Global Economic order (as illustrated, recently, by the regime of vaccine apartheid). According to the World Justice Project, which quantifies the "rule of law", we are currently in a rule of law recession—with the rule of law declining in most countries for the fifth year in a row in 2022.[1]

The law is identified as the antidote to most of the globe's maladies, including those on the list above. Yet we know remarkably little about the mechanics of how it works. What are the psychological processes and sociological dynamics involved in establishing—and sustaining—law? What is involved in an individual agent "opting in" to the law or acquiring what the legal philosopher HLA Hart referred to as the "internal point of view". This chapter, linking the jurisprudential literature on the subject with advances in moral psychology and the behavioural, cognitive and affective sciences, deepens the inquiry into the question of what constitutes "the internal point of view". Hume referred to psychology as "experimental philosophy"; in that tradition, this chapter locates itself in the emerging field of "experimental jurisprudence" (Sommers, 2021; Knobe & Shapiro, 2021; Tobia, 2022)—forging a link between legal theory and the frontiers of psychology.

---

[1] https://worldjusticeproject.org/index/

Legal Theory: From Sanction to Obligation

In mainstream jurisprudential discourses, it is now widely accepted that law is not merely an "external phenomenon", i.e., in the sense of being an entity or institution that can be understood from the perspective of observation alone, but, rather, that the adoption of the "internal perspective" is critical to its character. This means that to understand the law fully, it must be viewed from not just the outside but the inside.

The theories of John Austin (1832), Oliver Wendall Holmes (1897) and even Hans Kelsen (1934) were focused on law as sanctions, a perspective reiterated more recently by Frederick Schauer (2015). HLA Hart (1961)—arguably the most influential legal philosopher in the Anglo-American tradition—made a significant departure from this position. To rehash, briefly, the well-known dimensions of Hart's theory, he characterised a legal system as a union of primary and secondary rules (primary rules regulate behaviour directly, they are the rules *per se*; while secondary rules regulate the nature and application of primary rules, they are "rules about rules"). The existence of a legal system is, according to Hart, a matter of "social fact": the evidence for this lies in the observation that rules are generally obeyed, but critically also demands a dimension of active acceptance—secondary rules must be accepted, at least, by officials. Hart, thus, distinguishes the notion of obedience from that of acceptance.

Indeed, Hart insisted that the law requires participants of the social practice to assume what he calls the "internal point of view". According to Stephen Perry (1995), this implies that "an adequate jurisprudential account of law must at some point take into consideration how the practice looks to at least some of the practice's participants from the inside". Or, as Gerry Postema (1998) puts it, "The law, like other similar social practices, is constituted not only by intricate patterns of behavioral interaction, but beliefs, attitudes, judgements, and understandings of participants. The practice has an 'inside', the 'internal point of view' of participants". Scott Shapiro (2006) sums this up succinctly: the "internal point of view" is "a methodological prescription that demands that legal theories resonate with the shared experiences of *legal natives* [emphasis added]". This position is now standard among Anglo-American legal theorists.

In making this shift, Hart, critically, brings *participants* into the narrative of jurisprudence. The essence of this is captured by the distinction that he draws between law as a "sign" versus a "signal": a traffic light turning red does not merely *correlate* with cars on the road stopping, but it is the *cause* for cars to stop. Similarly, Hart draws a distinction between "being obliged" and "having an obligation"; the law is not simply a *prediction* of compliance, it is a *reason* for it. In other words, the law is, as the legal theorist Joseph Raz (1979) put it, a "reason for action". The hermeneutic—or interpretive—dimension is indispensable. The law is fundamentally normative and requires its members to congeal into a type of "institutional tribe".

The "internal point of view" is understood to be an essential feature of the law and appears to be doing a great deal of psychological heavy lifting from the perspective of the individual agent. But what *is* the "internal point of view" that is constitutive of this all-important institution? The conventional candidates in the jurisprudential discourse are—broadly—morality (the position broadly adopted by natural lawyers, often relying on law's divine origins),[2] coercion (the stance chosen by traditional legal positivists like Austin, with a cost-benefit variation being fielded by law and economics scholars like Ronald Coase) or, simply, habit, i.e., I obey the law because I think that it is the right thing to do or because I fear the consequences of disobedience or, simply because it has become a behavioural norm.

So, which of the three is it? Hart, himself, was attempting to set his position apart from theorists focused solely on sanctions, but what about morality? Shapiro (2006), attempting to ring-fence the "internal point of view" from morality, defines it as "a practical attitude of rule acceptance" and the product of "internalisation" on the part of the agent or participant. It is unclear—in psychological terms—what "acceptance" means (Shapiro excludes from this category both the theoretical observer tracking frequency of compliance, and the tactical "bad man"), or, for that matter, how the process of "internalisation" occurs. According to Hart, to "accept a social rule" means to treat it as "a general standard to be followed by the group as a whole" or as a "reason and justification" for action—but this is not necessarily synonymous with moral legitimacy (i.e., moral fact is clearly distinguished from social fact):

> allegiance to the system may be based on many different considerations: long term self-interest, disinterested interest in others; unreflecting inherited or traditional attitude; or the mere wish to do as others do. (Hart, 1961, p. 203)

Richard Holton (1998) puts forward a convincing argument to the contrary. He makes the case that Hart's own account of why the "internal point of view" is not a moral position is defective: "How can one both think that one morally ought not to accept the law, and yet still think that its demands are justified?" As he demonstrates, this position is untenable: it requires that Hart must show that "the right thing" to do is not necessarily the morally right thing, but what possible normative motivation could exist that is not moral? Holton equates the "internal point of view" with what he calls the moral attitude constraint and hopes to recast legal positivism as moral attitude positivism (salvaging it from accusations of both inconsistency and schizophrenia). Taking the view of law as a cluster concept, then, an official may not believe each individual law to be moral but at least that a systemic moral justification exists: a judge, for instance, justifying obedience to a law, is, by implication, upholding the morality of the law.

The theory suggests a variety of different psychological accounts of what underpins the law. It is, however, clear that Hart's definition demands a *degree* of psychological engagement of participants. There are two aspects to this: conformity in behaviour (the *sociological* aspect) and critical evaluation (the *psychological* aspect), i.e., criticising both self and other for non-conformity. While the

[2] See, as a reformulation of Aquinas, Finnis (1980)

behavioural dimension is the necessary condition, the normative component—invocations of "ought", "must" and "right" and "wrong", supported by "internal statements"—is the sufficient one. This sets laws, as social rules, apart from the "mere habits" of Austin's emphasis: aside from a high degree of compliance or an observable behavioural regularity (the behavioural), there is also the element of *psychological subscription* on the part of an individual agent or a critical reflective attitude (the attitudinal).

Hart, further, describes the existence of the "internal point of view" as a standing disposition: but how does this standing disposition come to be instilled or adopted? How is it sustained, and how can we ascertain when this standing disposition is critically at risk? What is the psychological story? How does the social rule that is law come to be a "behavioural regularity", i.e., which requires, for Hart, that most, if not every, member of a group engage in the behaviour? How does the "sovereign"—in modern societies, this is the "rule of law"—emerge? What are the sociological mechanics?

The internal point of view turns out to be critical for the very existence of a legal system—even Hart's model insists that at a very minimum what he refers to as "officials" must adopt it (a problematic proposition, in terms of setting the bar excessively low). But how can we look under the hood of this mysterious—but all-important—phenomenon, i.e., the social fact of the existence of law?

A Global Puzzle: Legal Transplantation

One possible way of starting to get a handle on these questions, empirically, would be to contrast contexts where sociological observation indicates that a legal system (and the attendant "internal point of view" that axiomatically underpins it) is present with those in which its presence is less certain, and to try to work out what is different in the two scenarios. An obvious global instance of this readily presents itself: ever since the decolonisation wave, starting in roughly 1945, a persistent puzzle for the post—World War has been the differential performance of legal systems in the Global North and Global South. This is captured by several metrics, including a range of Rule of Law indexes like the one produced by the World Justice Project (WJP).[3] These indicators are problematic in a variety of respects but provide at least a rough sketch of the global disparities in the functioning of the rule of law.[4] Strikingly, in its 2022 rankings, the top 50 spots in its list of 140 are occupied overwhelmingly by the nations of the Global North, while the bottom 90 are almost all nations of the Global South. Notably, Afghanistan ranks 138th or the second from the bottom, even after a multi-million dollar and 20-year long legal reform offensive in the country led by the United States. Indeed, this pattern—of the disparities in

[3] https://worldjusticeproject.org/index/global

[4] For a critique of attempts to quantify the law, see, for instance, Pistor et al. (2009)

performance in rule of law metrics—persists even after 75-odd years of constitutional, legislative and judicial interventions in formerly colonised nations.

What might explain this phenomenon? Depending on the taxonomy one uses, the globe can be divided into a relatively small number of "legal families". The French legal theorist Rene David (1950), writing in the 1950s and 1960s, identified *five* major categories: the Western tradition (which could further be divided into French or Romano-Germanic and Anglo-American Common Law), the Socialist, Islamic, Hindu and Chinese. The comparativists Zweigert et al. (1987) categorised *seven*: Romanistic, Germanic, Anglo-American, Nordic, Far Eastern, Islamic and Hindu. A more recent list by Berkowitz et al. (2003) boils most modern legal systems down to *four* distinct traditions: French, German, English and Scandinavian. There are other axes that have been treated by different authors as salient, for instance, there is an influential and voluminous literature on "legal origins" that attributes many aspects of economic performance to whether a nation has a common or a civil law system (Glaeser & Shleifer, 2002). For present purposes, however, the salient factor is that there are *very few* "originator" nations (all located in the Global North), while a *vast majority* of nations have been "recipients" of these systems (typically, as a function of patterns of colonisation). This broader phenomenon is referred to, in the literature, as "legal transplantation".

The term "legal transplant" was coined by the legal historian Alan Watson (1974) to describe "the moving of a rule [or legal system] from one country to another, or from one people to another"; literally to "transport to another country or place of residence" or shift to a place which is not native. Watson thought of legal rules as inherently nomadic and thought of "the transplanting of legal rules as socially easy"—at least in the Western world.[5] The jurist Pierre Legrand (1997), on the other hand, argues that the "historical, epistemological or cultural baggage" of rules is inherently dependent on the "interpretive community", rendering "legal transplants…impossible".[6] He argues that a meaningful "legal transplant" is only possible "when both the propositional statement as such *and* its invested meaning—which jointly constitute *the rule*—are transported from one culture to another", but that "as the words cross boundaries there intervenes a different rationality and morality to underwrite and effectuate the borrowed words".[7]

The empirical reality appears to lie somewhere in between these two positions. Using the examples of the "export" of English Common Law from Britain to its colonies in the "New World" (the United States, Canada, Australia and New Zealand) and India, Pakistan or Bangladesh, for instance—an example made famous by Acemoglu, Johnson and Robinson's work on the "reversal of fortunes"—the differential in performance is dramatic (Acemoglu et al., 2002). To return to the WJP 2022 rankings, as a ballpark: New Zealand ranks seventh, Canada ranks 12th, Australia ranks 13th and the United States ranks 26th while Pakistan ranks 129th,

---

[5] Watson (1974), p. 95

[6] p. 114–5

[7] ibid, p. 116–7

Bangladesh ranks 127th and India ranks 77th. If close to 200 countries around the world have inherited their legal systems from a handful of nations—half a dozen or so—it begs the question: why did transplants "take" in certain regions and not others? The answer, to build on Shapiro's analogy, may depend on how "legal natives" react to the "alien" system—and, as I will suggest in the next section, affective appeal.

Sticky Institutions: An Affective Phenomenon?

In this brief chapter, I want to present the hypothesis that the reason for the disappointing performance of legal reform efforts on a global scale may be a mismatch between the imported rules themselves (e.g. a methodologically individualistic rights-based view in an inherently collectivist culture) or other specific aspects of an externally imposed legal system (e.g. an excessive focus on procedure) and the psychological perspective, point of view or predisposition of agents native to a society—making it difficult for enough potential participants to fully adopt the "internal point of view" for a legal system to function, or at least function well.

What the "internal point of view" is in psychological terms remains an empirical mystery—and it has not yet been the subject of rigorous psychological testing. The theoretical positions on what it might be are wide-ranging: the law and economics school, for example, has largely reduced it to a cost-benefit analysis (thereby dismantling any normative component that it may have) while natural lawyers tend to equate it with morality (an explanation that is non-trivial in secular settings). But to the extent that it has been explicated, it has largely been assumed to be cognitive or rational. The American jurist Christopher Columbus Langdell, for instance, saw the law as the triumph of reason (Forsyth 2019).

Once a legal system takes root—either in its native, or foreign, soil—there is no doubt that the "rule of law" is singularly powerful in achieving rational social ends. It is, as popularly touted, one of the greatest triumphs of Western civilisation: it enables things to happen that would otherwise be unthinkable—from warring "tribes" of various stripes living in peace to anonymous economic transactions taking place between far corners of a country or beyond on digital portals, for example, like Amazon. However, the question at hand is that of the "buy in" of an individual agent *at the point* of transplantation. At this critical juncture, my argument is that it may be "affect" (broadly synonymous with emotion) rather than "cognition" that may need to be engaged. Institutions—of which law is the paradigmatic instance—have been described as "sticky": my contention is that human emotion may be the cause of this stickiness.

A little theoretical background first: the debate about the primacy of cognition versus emotion goes back to the 1800s. Darwin's seminal *The Expression of Emotion in Man and Animals* in 1872 made the case that emotions were mental states causing stereotypic bodily expressions, while William James' 1884 article "What Is Emotion?" contended that it is bodily activity that produces emotion, rather than vice versa. A version of the exchange is rehashed in the next century by Robert Zajonc and Richard Lazarus in their interchange on "feeling" versus "thinking". Zajonc's claim was that the idea of affect as a post-cognitive phenomenon sat uncomfortably with a variety of experimental results and clinical phenomena (Zajonc, 1984).[8] In many cases, including in decision-making, affect appears to precede cognition (with justifications often being produced *ex post*). In Zajonc's model, how an affective response interacts with the cognitive depends on, among other things, the strength of the affective response. In reply to Zajonc, Richard Lazarus (1984) emphatically rejected the notion of the independence of affect and dismissed the question of its primacy. He argued that "emotions reflect a constantly-changing person environment relationship". Thus, there is an external component inherent in emotional experience, and the internal experience can only be *understood* as emotion in the light of cognitive appraisal—by giving it "meaning or significance" (being startled does not, for instance, for Lazarus constitute an emotion). This, according to Lazarus, is the source of the intensity and quality of emotions generated. Even Lazarus readily admits, however, that "(e)motions appear to be powerful influences on how we think and interpret events. They are the result of cognition but in turn affect cognition". While the dichotomy between the two positions has been exaggerated, the empirical experience of legal transplantation seems to provide tangential evidence for Zajonc's view.

The legacy of this longstanding distinction between emotion and cognition is captured by Daniel Kahneman's highly influential work on the "slow brain" versus the "fast brain"—or the System 1 (instantaneous and automatic) and System 2 (deliberate) distinction—which has already made its way into the legal discourse (Kahneman, 2011). A variety of explanations exist for the causes for the development of the "fast" brain: one evolutionary account holds that our affective responses are on a continuum from non-human animals and derive from split-second responses that emerged as existential requirements of survival (for example, escaping a predator), i.e., that they "allow us to respond to environmental events rapidly and in a coordinated way" (Tsai & Colbert, 2019). Indeed, it is this psychological insight that the behavioural law and economics school, particularly the work by Thaler and Sunstein (2008) on *Nudge*, incorporate into legal reasoning by way of "bounded rationality" and "bounded willpower". Their central focus, however, is to mitigate the effects of deviance from pure rationality. As I have argued at length elsewhere though, this does not go nearly far enough: Thaler and Sunstein (2008), influenced by Kahneman, have successfully made a crack in the epistemic consensus—that door now needs to be flung open to accommodate a much wider range of psychological factors (Haldar, 2018a, b).

[8] Zajonc produced a list of the characteristics of affect: primary, basic, inescapable, irrevocable (in the sense of representing a "true" internal state or condition that, once experienced, cannot be rendered subjectively false), focused on the self (cognitive judgements deal with qualities that reside in the stimulus, while affective judgements are almost always to do with the self), non-verbal and prelinguistic, not dependent on cognition, and potentially separated from content.

Taking Affect Seriously

In this section, I would like to explore how taking the psychological make-up of actors generally, and "affect" specifically, more seriously could not only help solve an abiding intellectual and institutional puzzle—but also change our understanding of the working of the law.

Markus and Kitayama (1994) in their seminal work in the domain of cultural psychology emphasise that the priors to do with "goodness" or "morality" in different societies—and their accompanying practices—are *fundamentally* different. Variation ranges from certain societies being more individualistic and others more collectivist to some being "flatter" and others more "hierarchical"; from an emphasis on "independence" versus "interdependence"; from a preoccupation with "feeling good" or "having fun" or happiness to communal harmony; from the relative importance accorded to "positive" versus "negative" affect. What they call a "collective emotional reality" is highly salient.

Indeed, Markus and Kitayama (1991) establish that national differences on the individualism-collectivism axis produce different "models of the self"—"independent" (focused on distinctiveness, prioritising themselves, wanting to exert an influence to shape the environment) and "interdependent" (focused on connection to the point of being defined by duties and relationships, encouraged to prioritise others and applauded for adjusting to their environment). The outcome of this is that the most intense experiences are evoked for Westerners by "uniqueness" and for East Asians in the context of "connectedness". This distinction results, among other things, in widely divergent notions of the optimal amount of agency. Research by Tsai and Clobert (2019) confirms that "the dominant model of emotion in many East Asian contexts differs from that of many Western contexts". This is the empirically best-researched domain of these differences (for both ethnographic and theoretical reasons) but by no means the only one. They find, for instance, high cultural variation in what they refer to as "ideal affect" or the most desired emotional state—that Westerners favour high arousal and positive states, while Easterners prefer low arousal states (e.g. calmness) and have less of an aversion to negative states.

These findings appear to support the idea that at least one of the factors at play in contributing to the failure of law in much of the Global South may be its unsuitability to the cognitive/affective and cultural/contextual environment in which the institution attempts to take root. A few examples may be illustrative here. The "rights" paradigm, one of the shining beacons of the Western rule of law tradition, is fundamentally individualistic: it is focused on the vindication of the rights of well-informed and highly assertive individual actors and serves, at least in the immediate sense, to aid only their personal self-interest. In many societies, for example, East and South Asia, this would be actively frowned upon—especially for certain groups, like women. Another dimension of this is the insistence in the Western lexicon on procedural justice, for instance, being given notice in writing. This is a core dimension of "due process" in many Western societies where levels of both literacy and education are much higher, but, in many contexts, this step is simply meaningless.

The microfinance model, an informal lending scheme (there are no formal contracts between the bank and borrower, and loans are not collateralised) that I have studied closely, has been taken to task for its breach of due process in not giving borrowers written notice of failures to keep up with payments: but, in rural Bangladesh, where the model was first tested, the literacy levels of the borrowers (typically uneducated women) would have rendered this step largely meaningless (Haldar & Stiglitz, 2013).

An Emerging Field?

In his essay "The Emotional Dog and Its Rational Tail", Jonathan Haidt (2001) notes that it was only in the eighteenth century that English and Scottish philosophers started to talk about moral sentiments (a built-in moral sense that took pleasure in virtue and displeasure in vice) as an alternative to rationalism. This was true of Adam Smith—in *The Theory of Moral Sentiments*, the precursor to the better-known *The Wealth of Nations*—and, conspicuously, in the scholarship of David Hume. Hume believed that moral judgement emanated from sentiment rather than reason: "Reason is…the slave of the passions".

In presenting his social intuitionist model, Haidt makes the case that "moral judgement is caused by quick moral intuitions, and is followed (when needed) by slow, ex-post facto moral reasoning".[9] The four steps in the model are intuitive judgement, post hoc reasoning, direct persuasion and social persuasion. Haidt defines moral judgements as "evaluations (good vs. bad) of the actions or character of a person that are made with respect to a set of virtues held by a culture or subculture to be obligatory". Moral reasoning is defined as "conscious mental activity that consists of transforming given information…in order to reach a moral judgment". Moral intuition is defined as "the sudden appearance…of a moral judgment…without any conscious awareness of having gone through steps of search, weighing evidence, or inferring a conclusion".[10] It is important to note that the distinction between intuition and reasoning is different from that between emotion and cognition. According to Haidt, "morality, like language, is a major evolutionary adaptation for an intensely social species, built into multiple regions of the brain and body, which is better described as emergent than learned, yet which requires input and shaping from a particular culture".[11] Morality is, then, simultaneously innate and enculturated.

If this is indeed the case, the record of legal transplantation is unsurprising. Legal systems are currently failing to engage with agents at either the cognitive (including affective) or the cultural levels, providing a potential explanation for why these systems have not only failed to take root in many regions—but also why they are

[9] Haidt (2001), p. 5

[10] ibid, p. 6

[11] ibid, p. 17–8

undergoing a crisis of legitimacy even in places where they are more established. "Emotions, feelings, motivations, moods and other affective processes" are becoming increasingly central in a number of domains, so much so that a recent paper, circa 2021, asks if we are witnessing the era of "[t]he rise of affectivism" (as opposed to behaviourism or cognitivism). The change in the tide is attributable to a number of factors: we now know, for instance, that emotions are not always irrational (Frank, 1988), that reasoning is far from infallible (Kahneman & Tversky, 1984) and animals are not amoral (de Waal, 1996). At the same time, functional MRI images, and other methodological advances, have established human emotions as "objective, measurable, and scientifically observable phenomenon".[12]

As Terry Maroney, one of the scholars spearheading work on law and emotions, puts it, "the traditional narrative casting law as a bastion of pure reason has begun to crumble".[13] In 2015, in her book *Political Emotions*, the philosopher Martha Nussbaum (2015) made the case for how love was essential for the liberal project. I hope to have made a similar case for the importance of emotion for the legal one. The time is ripe for serious "law and emotions" scholarship.

References

Acemoglu, D., Johnson, S., & Robinson, J. A. (2002). Reversal of fortune: Geography and institutions in the making of the modern world income distribution. *The Quarterly Journal of Economics, 117*(4), 1231–1294.

Austin, J. L. (1968 [1832]). *The providence of jurisprudence determined; and the uses of the study of jurisprudence*. Weidenfeld and Nicolson.

Berkowitz, D., Pistor, K., & Richard, J. F. (2003). The transplant effect. *American Journal of Comparative Law, 51*(1), 163.

Coase, R. H. (1988). *The firm, the market and the law/R.H. Coase*. University of Chicago Press.

Darwin, C. (1998 [1872]). *The expression of emotion in man and animals*. HarperCollins.

David, René (1950): Traité élémentaire de droit civil comparé 225.

de Waal, F. (1996). *Good natured: The origins of right and wrong in humans and other animals*. Harvard University Press.

Dukes, D., et al. (2021). The rise of affectivism. *Nature Human Behaviour, 5*, 816–820.

Finnis, J. (1980). *Natural law, natural rights*. Clarendon Press.

Forsyth, A. (2019). Law as science. In *Common law and natural law in America*. Cambridge University Press.

Frank, R. (1988). *Passions within reason: The strategic role of the emotions*. W. W. Norton.

Glaeser, E. L., & Shleifer, A. (2002). Legal origins. *Quarterly Journal of Economics, 117*(4), 1193–1229.

Haidt, J. (2001). The emotional dog and its rational tail. *Psychological Review, 108*(4), 814–834.

Haldar, A. (2018a, November 2). Intrinsic goodness. *Times Literary Supplement.*

Haldar, A. (2018b, December 14). Economics: The discipline that refuses to change. *The Atlantic.*

Haldar, A. (2022a, February 15). Boris Johnson's last affair. *Project Syndicate*, https://www.project-syndicate.org/commentary/boris-johnson-partygate-scandal-by-antara-haldar-2022-02

[12] Dukes et al. (2021)

[13] Maroney (2016)

Haldar, A (2022b, April 18). Why do we obey the law? *Project Syndicate*, https://www.project-syndicate.org/commentary/ketanji-brown-jackson-bolsters-supreme-court-sagging-legitimacy-by-antara-haldar-2022-04
Haldar, A. (2022c, May 18). Losing rights in America. *Project Syndicate*, https://www.project-syndicate.org/commentary/supreme-court-reversing-womens-rights-by-antara-haldar-2022-05
Haldar, A. (2022d, June 29). Trump's lessons for defending the rule of law. *Project Syndicate*, https://www.project-syndicate.org/commentary/us-january-6-committee-trump-facts-emotion-by-antara-haldar-2022-06
Haldar, A. (2022e, October 26). Libertarian lettuce. *Project Syndicate*, https://www.project-syndicate.org/commentary/truss-uk-libertarian-experiment-failed-by-antara-haldar-2022-10
Haldar, A., & Stiglitz, J. (2013). Analyzing legal formality and informality. In D. Kennedy & J. Stiglitz (Eds.), *Law and economics with Chinese characteristics*. Oxford University Press.
Hart, H. L. A. (1961). *The concept of law*. Clarendon Press.
Holmes, O. W. (1897). The path of the law. *Harvard Law Review, 10*(8), 457–478.
Holton, R. (1998). Positivism and the internal point of view. *Law and Philosophy, 17*(5/6), 597–625.
Hume, D. (2000 [1739]). *A treatise of human nature*. Oxford University Press.
James, W. (1884). What is emotion? *Mind, 9*(34), 188–205.
Kahneman, D., & Tversky, A. (1984). Choices, values, and frames. *American Psychologist, 39*(4), 341–350.
Kahnemann, D. (2011). *Thinking fast and slow*. Farrar, Straus and Giroux.
Kelsen, H. (1934). Pure theory of law, the-its method and fundamental concepts. *Law Quarterly Review*, 50, 474.
Knobe, J., & Shapiro, S. (2021). Proximate cause explained: An essay in experimental jurisprudence. *The University of Chicago Law Review, 88*(1), 165.
Lazarus, R. (1984). On the primacy of cognition. *American Psychologist, 39*, 124–129.
Legrand, P. (1997). On the impossibility of "legal transplants". *Maastricht Journal of European and Comparative Law, 4*(2), 111–124.
Markus, H. R., & Kitayama, S. (1991). Culture and the self: Implications for cognition, emotion and motivation. *Psychological Review, 98*(2), 224–253.
Markus, H. R., & Kitayama, S. (1994). The cultural shaping of emotion: A conceptual framework. In S. Kitayama & H. R. Markus (Eds.), *Emotion and culture: Empirical studies of mutual influence* (pp. 339–351). American Psychological Association.
Maroney, T. (2016). A field evolves: Introduction to the special section on law and emotion. *Emotion Review, 8*(1), 3–7.
Nussbaum, M. (2015). *Political emotions: Why love is necessary for justice*. Harvard University Press.
Perry, S. R. (1995). Interpretation and methodology in legal theory. *Law and Interpretation, 97*(99), 188–189.
Pistor, K., Haldar, A., & Amirapu, A. (2009). Social norms, rule of law and gender reality. In J. J. Heckman, R. Nelson, & L. Cabbatingan (Eds.), *Global perspectives on the rule of law*. Routledge-Cavendish.
Postema, G. J. (1998). Jurisprudence as practical philosophy. *Legal Theory, 329*, 332.
Raz, J. (1979). *The authority of law: Essays on law and morality*. Clarendon Press.
Schauer, F. (2015). *The force of law*. Harvard University Press.
Shapiro, S. (2006). What is the internal point of view? *Fordham Law Review, 75*(3), 1157.
Smith, A. (1776). An inquiry into the nature and causes of the wealth of nations: Volume one. :. Printed for W. Strahan; and T. Cadell, 1776.
Smith, A. *The theory of moral sentiments. By Adam Smith, Professor of Moral Philosophy in the University of Glasgow*. The second. Printed for A. Millar, in the Strand; and A. Kincaid and J. Bell, in Edinburgh, 1761. Print.
Sommers, R. (2021). Experimental jurisprudence. *Science, 373*(6553), 394–395.

Thaler, R., & Sunstein, C. (2008). *Nudge: Improving decisions about health, wealth and happiness*. Yale University Press.
Tobia, K. (2022). Experimental jurisprudence. *The University of Chicago Law Review, 89*(3), 735–802.
Tsai, J., & Clobert, M. (2019). Cultural influences on emotion: Empirical patterns and emerging trends. In S. Kitayama & D. Cohen (Eds.), *Handbook of cultural psychology* (pp. 292–318). Guilford Press.
Watson, A. (1974). *Legal transplants: An approach to comparative law*. Scottish Academic Press.
Zajonc, R. (1984). On the primacy of affect. *American Psychologist, 39*, 117–123.
Zweigert, K., Kötz, H., & Weir, J. A. (1987). *An introduction to comparative law* (2nd edn).

Antara Haldar is the inaugural Professor of Empirical Legal Studies at the University of Cambridge, with appointments at the Faculties of Law, Business, and Philosophy. She is also a Visiting Professor in the Department of Psychology at Harvard University and held a prestigious 2020–21 Fellowship at Stanford University's Center for Advanced Study in the Behavioral Sciences. She is the Principal Investigator of a multimillion-dollar European Research Council grant, LIMEN, and is popularly regarded as one of the world's top authorities on institutions and cognition. She is a regular contributor to some of the world's leading publications including *The Atlantic*, *WIRED*, the *Times Literary Supplement*, *The Independent*, the *Globe and the Mail*. She writes a widely read monthly column for *Project Syndicate* that applies her own cutting-edge research across disciplinary silos to the most pressing issues of the day. She is the author of over 200 scholarly and journalistic articles, and her work has already been translated into over 20 languages.

The Epistemic Argument for Hedonism

Neil Sinhababu

Abstract I defend ethical hedonism, the view that pleasure is the sole good thing, by arguing that it offers the only answer to an argument for moral skepticism. The skeptical problem arises from widespread fundamental moral disagreement, which entails the presence of enough moral error to undermine the reliability of most processes generating moral belief. We know that pleasure is good through the reliable process of phenomenal introspection, which reveals what our experiences are like. If knowing of pleasure's goodness through phenomenal introspection is the only reliable way to form our initial moral beliefs, we should accept ethical hedonism.

Keywords Pleasure · Hedonism · Reliability · Disagreement · Introspection

I argue that we should give up all of our moral beliefs, except for the belief in pleasure's goodness and whatever it entails about virtue and right action. This is a version of ethical hedonism on which we doubt rather than reject nonhedonic moral facts.[1] As I agree with the consensus that hedonism is counterintuitive, I won't argue that it best fits ordinary intuition.[2]

Instead, I present an argument for moral skepticism and show that hedonism has a unique answer to it. The first half of this chapter presents the skeptical argument, which arises from the breadth of moral disagreement. Widespread disagreement involves enough false moral belief to entail the unreliability of the processes by which we form moral beliefs, bringing moral skepticism upon us. The second half

[1] According to ethical hedonism, pleasure's goodness and whatever it entails are the only moral facts. Ethical hedonism combined with suitable aggregative and consequentialist principles entails hedonic utilitarianism. "Pleasure's goodness" here includes displeasure's badness.

[2] Many cases suggest that hedonism is counterintuitive – for example, the Experience Machine from Nozick (1974). Many other objections are canvassed in Chap. 3 of Feldman (2004).

N. Sinhababu (✉)
National University of Singapore, Kallipolis, Singapore

S. Chakraborty (ed.), *Human Minds and Cultures*,
https://doi.org/10.1007/978-3-031-56448-2_8

of this chapter argues that belief in pleasure's goodness escapes skeptical defeat. Pleasure's goodness can be known through the highly reliable process of phenomenal introspection. If introspecting pleasure's goodness is the only answer to the skeptical argument, our only moral beliefs should be in pleasure's goodness and whatever it entails.

The Argument from Disagreement [1]

This half of the chapter presents the *argument from disagreement*. 1–4 are premises:

1. In any moral disagreement, at least one party must be in error.
2. There is widespread moral disagreement.
3. If there is widespread error about a topic, we should retain only those beliefs about it formed through reliable processes.
4. If there is widespread error about morality, there are no actual reliable processes for forming moral beliefs.
5. There is widespread error about morality (from 1 and 2).
6. We should retain only those moral beliefs formed through reliable processes (from 3 and 5).
7. There are no actual reliable processes for forming moral beliefs (from 4 and 5).
8. We should give up all of our moral beliefs (from 6 and 7).

This section briefly explains how the argument works and distinguishes it from other recent arguments involving disagreement. The next three sections defend 2–4 at length.

1 lays out how moral disagreement must work – at least one party must be in error. The objectivity of morality entails this. Parties to moral disagreements assert genuinely contradictory propositions, some of which must be false. Noncognitivists and constructivists might deny 1, claiming that moral disagreement need not involve error on any side. I can't do their views justice here, and I assume that noncognitivist and constructivist construals of moral concepts are mistaken, and any genuine moral facts must be objective.[3]

2 refers only to disagreement about fundamental moral questions, not disagreement that arises from agreement on these questions combined with differing opinions on nonmoral matters. Disagreement arising from differing nonmoral beliefs doesn't advance the argument, as it doesn't bear on the reliability of our processes for forming moral beliefs. 1 and 2 imply 5, that there is widespread moral error. Since parties to a moral disagreement contradict each other about an objective fact, each disagreement involves false belief. Disagreement between many parties about many things involves many false beliefs about many things.

[3] Semantic problems with noncognitivism are explored by Schroeder (2008) and Merli (2008).

What should we do if there is widespread error about some topic, as 5 suggests? We should follow 3 and retain only those beliefs about the topic formed through reliable processes. The processes driving belief formation on the topic leave people at a high risk of having false beliefs. If we hold beliefs on the topic and the processes that created them are of average reliability for the topic, they run a high risk of being false. If this risk is high enough, we should abandon our beliefs. But we need not abandon our beliefs if an unusually reliable process of belief formation supports them.

While it's easy to see how a process responsible for a small fraction of our moral beliefs might be reliable amid widespread error, 4 says that none of the actual processes we rely on are reliable. I defend 4 by considering the processes of belief formation suggested in the philosophical and psychological literature, and arguing that widespread error makes them unreliable. (The second half of this chapter answers the skeptical argument by offering what I think is the only actual counterexample to 4 – phenomenal introspection, which reliably informs us of pleasure's moral value.) So defending 4 requires denying the reliability of every such minor process. I argue against many of the most-discussed contenders. 4 and 5 together imply 7, the absence of any reliable processes of belief formation. When 6, which only allows moral beliefs formed through reliable processes, is combined with 7, we arrive at 8 and must abandon all our moral beliefs.

The argument from disagreement avoids the difficulties with explanatory arguments from disagreement, like John Mackie's (1977) argument from relativity. Mackie argues that moral belief isn't best explained by objective moral value, since there are so many differing views. He writes that "the actual variations in the moral codes are more readily explained by the hypothesis that they reflect ways of life than by the hypothesis that they express perceptions, most of them seriously inadequate and badly distorted, of objective values" (37).[4] The argument from disagreement doesn't suggest rejecting morality for its explanatory inadequacy, as Mackie does. Explanatory arguments against morality might be implausible or self-undermining if they also force us to reject color, numbers, or epistemic norms, which are similarly hard to place in our best explanations.[5] The argument from disagreement avoids these problems, as it applies only when widespread disagreement entails so much error that our beliefs must be unreliably formed. There is less disagreement about color, mathematics, and epistemic norms.

Much recent work in epistemology discusses disagreement between epistemic peers, especially disagreement between two people about a single proposition.[6] As Katia Vavova (2014) argues, this work doesn't support moral skepticism.[7] But even if disagreement with one epistemic peer about one proposition doesn't force us to revise our beliefs, disagreement with the rest of humanity about entire theories can

---

[4] This is how Brink (1989) understands Mackie's argument.

[5] See Sayre-McCord (1988), responding to Harman (1988).

[6] For example, Kelly (2005) and Elga (2007).

[7] See also Decker and Groll (2013).

demonstrate the general unreliability of human belief-forming processes and make us revise our beliefs. The argument from disagreement does this, moving from widespread disagreement to widespread error to the unreliability of human belief formation to the unreliability of an individual's belief formation.[8]

Several other arguments from moral disagreement to skepticism or anti-realism are nicely criticized by Michael Huemer (2005) and David Enoch (2009). But Enoch and Huemer don't address arguments that use the presence of error in each moral disagreement to show that widespread moral disagreement entails widespread error, which shows that our processes of moral belief formation are unreliable. My argument is the only one that does this.

Richard Joyce (2001) and Sharon Street (2006) present evolutionary debunking arguments that can support moral skepticism, just as the argument from disagreement does. They argue that our moral beliefs result from evolutionary processes that are unlikely to have any systematic correlation with objective moral truth. Seeing this, we should abandon our beliefs in objective moral facts. This evolutionary debunking argument offers another route to the skeptical conclusion of the argument from disagreement. The adaptive benefits of knowing what one's experiences are like enable phenomenal introspection to answer evolutionary debunking arguments as well.

There Is Widespread Moral Disagreement [2]

This section argues that there is enough moral disagreement to drive the argument from disagreement forward. Conclusively demonstrating this would require more anthropological and historical work than any paper could contain. Perhaps a reasonable aim is to make the existence of that much moral disagreement plausible. First, I present some evidence that actual moral disagreement is this great. To answer the objection that differing opinions about moral questions really arise from nonmoral disagreement, I argue that the different nonmoral beliefs are often the products of underlying moral disagreement. Finally, I argue that the narrow areas of moral agreement remaining after taking this point into account won't block the argument from disagreement.

For a vivid illustration of how intensely we disagree about morality, consider a scene from the end of Homer's (1900) *Odyssey*. Odysseus has returned home and killed the 100 suitors who sought his wife's hand. He and his son Telemachus bring out 12 of the servant women and make them assist in cleaning up the dead bodies.

> Then when they had made the whole place quite clean and orderly, they took the women out and hemmed them in the narrow space between the wall of the domed room and that of the yard, so that they could not get away: and Telemachus said to the other two, "I shall not let

[8] For a helpful discussion of multiparty intertheoretic disagreement that inspires the argument presented here, see Stewart (2005).

> these women die a clean death, for they were insolent to me and my mother, and used to sleep with the suitors."
>
> So saying he made a ship's cable fast to one of the bearing-posts that supported the roof of the domed room, and secured it all around the building, at a good height, lest any of the women's feet should touch the ground; and as thrushes or doves beat against a net that has been set for them in a thicket just as they were getting to their nest, and a terrible fate awaits them, even so did the women have to put their heads in nooses one after the other and die most miserably. Their feet moved convulsively for a while, but not for very long.

The *Odyssey*'s treatment of these events demonstrates how dramatically ancient Greek moral intuitions differ from ours. It doesn't dwell on the brutality of Telemachus, who killed 12 women for the trivial reasons he states, making them suffer as they die. While gods and men seek vengeance for other great and small offenses in the *Odyssey*, no one finds this mass murder worth avenging. It's a minor event in the denouement to a happy ending in which Odysseus (who first proposes killing the women) returns home and Telemachus becomes a man. That the Greeks could so easily regard these murders as part of a happy ending for heroes shows how deeply we disagree with them. It's like a trolley problem with the 12 women on the side track and no one on the main track, in which they judged it permissible for Telemachus to turn the trolley to kill them all. And this isn't some esoteric text of a despised or short-lived sect, but a central literary work of a long-lived and influential culture.

Human history offers similarly striking examples of disagreement on a variety of topics. These include sexual morality; the treatment of animals; the treatment of other ethnicities, families, and social classes; the consumption of intoxicating substances; whether and how one may take vengeance; slavery; whether public celebrations are acceptable; and gender roles.[9] Moral obligations to commit genocide were accepted not only by some twentieth-century Germans but by much of the ancient world, including the culture that gave us the Old Testament. One can only view the human past and much of the present with horror at the depth of human moral error and the harm that has resulted.

One might think to explain away much of this disagreement as the result of differing nonmoral beliefs. Those who disagree about nonmoral issues may disagree on the moral rightness of a particular action despite agreeing on the fundamental moral issues. For example, they may agree that healing the sick is right, but disagree about whether a particular medicine will heal or harm. This disagreement about prescribing the medicine won't be fundamentally about morality, and won't support the argument from disagreement.

I don't think the moral disagreements listed above are explained by differences in nonmoral belief. This isn't because sexists, racists, and bigots share the nonmoral views of those enlightened by feminism and other egalitarian doctrines – they don't. Rather, their differing views on nonmoral topics often are rationalizations of moral

[9] See Doris and Plakias (2008).

beliefs that fundamentally disagree with ours.[10] Those whose fundamental moral judgments include commitments to the authority of men over women, or of one race over another, will easily accept descriptive psychological views attributing less intelligence or rationality to women or the subjugated race.[11] Moral disagreement supposedly arising from moral views in religious texts is similar. Given how rich and many-stranded most religious texts are, interpretive claims about their moral teachings often tell us more about the antecedent moral beliefs of the interpreter than about the text itself. This is why the same texts are interpreted to support so many different moral views. Similar phenomena occur with most moral beliefs. Environmentalists who value a lovely patch of wilderness will easily believe that its destruction will cause disaster, those who feel justified in eating meat will easily believe that the animals they eat don't suffer greatly, and libertarians who feel that redistributing wealth is unjust will easily believe that it harms the economy. We shouldn't assume that differing moral beliefs on practical questions result from fundamental moral agreement combined with differing nonmoral beliefs. Often, the differing nonmoral beliefs are caused by fundamental moral disagreement.

As we have no precise way of quantifying the breadth of disagreement or determining its epistemic consequences, it's unclear exactly how much disagreement the argument requires. While this makes the argument difficult to evaluate, it shouldn't stop us from proceeding, as we have to use the unclear notion of widespread disagreement in ordinary epistemic practice. If 99.9% of botanists agree on some issue about plants, non-botanists should defer to their authority and believe as most of them do. But if disagreement between botanists is suitably widespread, non-botanists should remain agnostic. A more precise and systematic account of when disagreement is widespread enough to generate particular epistemic consequences would be very helpful. Until we have one, we must employ the unclear notion of widespread disagreement.

Against the background of widespread moral disagreement, there may still be universal or near-universal agreement on some moral questions. For example, perhaps all cultures agree that one should provide for one's elderly parents, even though they generally disagree elsewhere. How do these narrow areas of moral agreement affect the argument?

This all depends on whether the narrow agreement is reliably or unreliably caused. If narrow agreement results from reliable processes of belief formation, it lets us avoid error, defeating the argument from disagreement. But widely accepted moral beliefs may result from widely prevailing unreliable processes leading everyone to the same errors. There's no special pressure to explain agreement in terms of reliable processes when disagreement is widespread. Explaining agreement in terms of reliable processes is preferable when we have some reason to think that the processes involved are generally reliable. Then we would want to understand cases of

[10] For a good discussion, see Arpaly (2003). For striking psychological evidence about rationalization, see Haidt (2001).

[11] As system justification theorists in psychology argue, even subjugated groups form beliefs according to which their subjugation is justified. See Jost et al. (2004).

agreement in line with the general reliability of processes producing moral belief. But if disagreement is widespread, error is too. Since moral beliefs are so often false, invoking unreliable processes to explain them is better than invoking reliable ones. The next two sections discuss this in more detail.

We have many plausible explanations of narrow agreement on which moral beliefs are unreliably caused. Evolutionary and sociological explanations of why particular moral beliefs are widely accepted often invoke unreliable mechanisms.[12] On these explanations, we agree because some moral beliefs were so important for reproductive fitness that natural selection made them innate in us, or so important to the interests controlling moral education in each culture that they were inculcated in everyone. For example, parents' influence over their children's moral education would explain the agreement that one should provide for one's elderly parents. Plausible normative ethical theories won't systematically connect these evolutionary and sociological explanations with moral facts. If disagreement and error are widespread, they'll provide useful ways to reconcile unusual cases of widespread agreement with the general unreliability of the processes producing moral belief.

If There Is Widespread Error About a Topic, We Should Retain Only Those Beliefs About It Formed Through Reliable Processes [3]

Now I defend 3. First, I show how the falsity of others' beliefs undermines one's own belief. Then I clarify the notion of a reliable process, and show how it's friendly to both epistemic internalists and externalists. I illustrate 3's plausibility by considering cases where it correctly guides our reasoning. Finally, I show how 3 is grounded in the intuitive response to grave moral error.

First, a simple objection: "Why should I care whether other people have false beliefs? That's a fact about other people, and not about me. Even if most people are wrong about some topic, I may be one of the few right ones, even if there's no apparent reason to think that my way of forming beliefs is any more reliable."

While widespread error leaves open the possibility that one has true beliefs, it reduces the probability that my beliefs are true. I have no direct evidence that I have an appendix, but I know that previous investigations have revealed appendixes in humans. So induction suggests that I have an appendix. Similarly, I know on the basis of 1 and 2 that human moral beliefs are, in general, rife with error. So even if I have no direct evidence of error in my moral beliefs, induction suggests that they are rife with error as well.

3 invokes the reliability of the processes that produce our beliefs. Assessing processes of belief formation for reliability is an important part of our epistemic

[12] For evolutionary explanations, see Street (2006) and Joyce (2001). For classic sociological explanations, see Foucault (1977), Nietzsche (1998).

practices. If I'm told that my belief is entirely produced by wishful thinking, I can't simply accept that and maintain the belief. Knowing that wishful thinking is unreliable, I must either deny that my belief is entirely caused by wishful thinking or abandon the belief. But if someone tells me that my belief is entirely the result of visual perception, I'll maintain it, assuming that it concerns sizable nearby objects or something else about which visual perception is reliable. While providing precise criteria for individuating processes of belief formation is hard, as the literature on the generality problem for reliabilism attests, individuating them somehow is indispensable to our epistemic practices.[13] Following Alvin Goldman's (2000) remark that "It is clear that our ordinary thought about process types slices them broadly" (346), I treat cognitive process types like wishful thinking and visual perception as appropriately broad. Trusting particular people and texts, meanwhile, is too narrow. Cognitive science may eventually help us better individuate cognitive process types for the purposes of reliability assessments and discover which processes produce which beliefs.

Epistemic internalists might reject 3 as stated, claiming that it isn't widespread error that would justify giving up our beliefs, but our having reason to believe that there is widespread error. They might also claim that our justification for believing the outputs of some process depends not on its reliability, but on what we have reason to believe about its reliability. 3 can be modified to suit internalist tastes, changing its antecedent to "If we have reason to believe that there is widespread error about a topic" or changing its consequent to "we should retain only those beliefs about it that we have reason to believe were formed through reliable processes." While 3's antecedent might itself seem unnecessary on the original formulation, it's required for 3 to remain plausible on the internalist modification. Requiring us to have reason to believe that any of our belief formation processes are reliable before retaining their outputs might lead to skepticism. The antecedent limits the scope of the requirement to cases of widespread error, averting general skeptical conclusions. The argument will still attain its conclusion under these modifications. Successfully defending the premises of the argument and deriving widespread error (5) and unreliability (7) gives those of us who have heard the defense and derivation reason to believe 5 and 7. This allows us to derive 8. (Thus the pronoun "we" in 3, 6, and 8.)

3 describes the right response to widespread error in many actual cases. Someone in the twelfth century, especially upon hearing the disagreeing views of many cultures regarding the origins of the universe, would do well to recognize that error on this topic was widespread and retreat to agnosticism about it. Only when modern astrophysics extended reliable empirical methods to cosmology would it be rational to move forward from agnosticism and accept a particular account of how the universe began. Similarly, disagreement about which stocks will perform better than average is widespread among investors, suggesting that one's beliefs on the matter

[13] See Conee and Feldman (1998). The generality problem seeems to apply to all plausible epistemological views, as argued by Comesaña (2006).

have a high likelihood of error. It's wise to remain agnostic about the stock market without an unusually reliable way of forming beliefs – for example, the sort of secret insider information that it's illegal to trade on.

3 permits us to retain our moral beliefs in individual cases of moral disagreement, suggesting skeptical conclusions only when moral disagreement is widespread. When we consider a single culture's abhorrent moral views, like the Greeks' acceptance of Telemachus and Odysseus' murders of the servant women, we don't think that maybe the Greeks were right to see nothing wrong and we should reconsider our outrage. Instead, we're horrified by their grave moral error. I think this is right. We're similarly horrified by the moral errors of countless others, including Hindus who burned widows on their husbands' funeral pyres, American Southerners who supported slavery and segregation, and our contemporaries who condemn homosexuality. The sheer number of cases like this requires us to regard moral error as a pervasive feature of the human condition. Humans typically form moral beliefs through unreliable processes and have appendixes. We are humans, so this should reduce our confidence in our moral judgments. The prevalence of error in a world full of moral disagreement demonstrates how bad humans are at forming true moral beliefs, undermining our own moral beliefs. Knowing that human moral beliefs are so often unreliably caused, we'll require our moral beliefs to issue from reliable processes.

If There Is Widespread Error About Morality, There Are No Actual Reliable Processes for Forming Moral Beliefs [4]

A reliable process for forming moral beliefs would avert skeptical conclusions. I consider the processes widely taken to produce our moral beliefs and argue that they don't help us escape moral skepticism. Ordinary moral intuition, whether it involves a special rational faculty or our emotional responses, is shown to be unreliable by the existence of widespread error. The argument from disagreement either prevents reflective equilibrium from generating moral conclusions or undermines it. Conceptual analysis is reliable, but delivers the wrong kind of knowledge to avert skepticism. If all our processes for forming moral beliefs are unreliable, moral skepticism looms. 4 is false only because of one process – phenomenal introspection, which lets us know of the goodness of pleasure, as the second half of this chapter discusses.

Any general process of moral intuition must be unreliable, because widespread error guarantees the unreliability of any process by which we form all or almost all of our moral beliefs. This is why the antecedent of 4 is important. While widespread error allows some processes responsible for a small share of our moral beliefs to be reliable, it implies that any process generating most or all moral beliefs must be unreliable. Since the process produced so many of our moral beliefs, and so many of them are erroneous, it must be responsible for a large share of the error. If more of people's moral beliefs were true, things would be otherwise. Widespread truth would support the reliability of any process that produced most or all of our moral beliefs, since that process would be responsible for so much true belief. But given widespread error, ordinary moral intuition must be unreliable. This provides a forceful response to Moorean opponents who insist that we can't give up the reliability of a process by which we form all or nearly all of our beliefs on an important topic. Even if this Moorean response helps against external world skeptics who point to widespread error among non-actual brains in vats, it doesn't help against moral skeptics who use 1 and 2 to derive widespread actual error – something the external world skeptic could never demonstrate. I briefly consider two conceptions of moral intuition – as a special rational faculty by which we grasp nonnatural moral facts, and as a process by which our emotions lead us to form moral beliefs – and show how widespread error guarantees their unreliability.

Nonnaturalists like Huemer and Enoch regard moral intuition as involving a special rational faculty that lets us know nonnatural moral facts. They argue that this rational faculty produces knowledge on many topics including mathematics, logic, and modality, so moral knowledge might operate similarly. They might defend the reliability of moral intuition in the face of widespread error by arguing that if intuition is reliable about these other things, its overall reliability across moral and nonmoral areas allows us to reliably form moral beliefs by using it. This defense won't work. When an epistemic process is manifestly unreliable on some topic, as widespread error shows any process responsible for most of our moral beliefs to be, the reliability of that process elsewhere won't save it on that topic. The general reliability of testimony doesn't imply that compulsive gamblers' testimony is reliable about the next spin of the roulette wheel. Even if intuition is reliable elsewhere, widespread moral disagreement renders it unreliable in ethics.

Others regard ordinary moral intuition as driven by emotional responses which cause moral beliefs. Just as visual experiences of color cause beliefs about the colors of things, emotional experiences cause beliefs about the moral properties of things.[14] Feeling guilt about one's action causes the belief that one acted wrongly, admiring another person causes the belief that that person is virtuous, and horror at a state of affairs causes the belief that it is bad. We might have regarded this as a reliable way to know about moral facts, just as visual perception is a reliable way to know about color, if not for widespread error. But because of widespread error, we can only see it as an unreliable process responsible for our dismal epistemic situation.

Reflective equilibrium is the prevailing methodology in normative ethics today. It involves changing our beliefs about particular cases and general principles to make them cohere. Whether or not nonmoral propositions like the premises of the argument from disagreement are admissible in reflective equilibrium, widespread error prevents reflective equilibrium from reliably generating a true moral theory, as I explain.

---

[14] Schnall (2008). For related views, see Kauppinen (2013), Werner (2016), Tappolet (2016), Sinhababu (2017).

If the premises of the argument from disagreement are admitted into reflective equilibrium, the argument can be reconstructed there, and reflective equilibrium will dictate that we give up all of our moral beliefs. To avoid this conclusion, the premises of the argument from disagreement would have to be revised away on moral grounds. These premises are a metaethical claim about the objectivity of morality which seems to be a conceptual truth, an anthropological claim about the existence of disagreement, a very general epistemic claim about when we should revise our beliefs, and a more empirically grounded epistemic claim about our processes of belief formation and their reliability. While reflective equilibrium may move us to revise substantive moral beliefs in view of other substantive moral beliefs, claims of these other kinds are less amenable to such revision. Unless ambitious arguments for revising these nonmoral claims away succeed, we must follow the argument to its conclusion and accept that reflective equilibrium makes moral skeptics of us.[15]

If only moral principles and judgments are considered in reflective equilibrium, it won't make moral skeptics of us, but the argument from disagreement will undermine its conclusions. The argument forces us to give up the preexisting moral beliefs against which we test various moral propositions in reflective equilibrium. While we may be justified in believing something because it coheres with our other beliefs, this justification goes away once we see that those beliefs should be abandoned. Coherence with beliefs known to be unjustified doesn't confer justification.

Now I consider conceptual analysis. It can produce moral beliefs about conceptual truths – for example, that the moral supervenes on the nonmoral, and that morality is objective. It also may provide judgments about relations between different moral concepts – perhaps, that if the only moral difference between two actions is that one would produce morally better consequences than the other, doing what produces better consequences is right. I regard conceptual analysis as reliable, so that the argument from disagreement doesn't force us to give up the beliefs about morality it produces. Unfortunately, if analytic naturalism is false, as has been widely held in metaethics since G. E. Moore (1903), conceptual analysis won't provide the knowledge we need to start building a normative ethical theory.[16] Even if it tells us that producing better states of affairs contributes to an action's rightness, it doesn't tell us that anything is good or right to begin with. That's the knowledge we need to avoid moral skepticism.

So far I've argued that we're forced to abandon our moral beliefs by our epistemic and anthropological situation, when combined with reasonable metaethical and epistemic principles. But if a reliable process of moral belief formation exists, 4 and moral skepticism are false. The rest of this chapter discusses the only reliable process I know of.

[15] For a discussion of how odd it would be to revise away empirical nonmoral claims on moral grounds, see Barber (2013).

[16] For dissent, see Jackson (1998). Even if Jackson is right, moral concepts may be complex enough that epistemic difficulties arise again.

Phenomenal Introspection Reveals Pleasure's Goodness [5]

Phenomenal introspection, a reliable way of forming true beliefs about our experiences, produces the belief that pleasure is good. Even as our other processes of moral belief formation prove unreliable, it provides reliable access to pleasure's goodness, justifying the positive claims of hedonism. This section clarifies what phenomenal introspection and pleasure are and explains how phenomenal introspection provides reliable access to pleasure's value. The next section argues that pleasure's goodness is genuine moral value, rather than value of some other kind.

In phenomenal introspection, we consider our subjective experience, or phenomenology, and determine what it's like. Phenomenal introspection can be reliable while dreaming or hallucinating, as long as we can determine what the dreams or hallucinations are like. By itself, phenomenal introspection doesn't produce beliefs about things outside experience, or about relations between our experiences and non-experiential things. So it doesn't produce judgments about the rightness of actions or the goodness of non-experiential things. It can only tell us about the intrinsic properties of experience itself.

Phenomenal introspection is generally reliable, even if mistakes about immediate experience are possible. Experience is rich in detail, so one could get some of the details wrong in belief. Under adverse conditions involving false expectations, misleading evidence about what one's experiences will be, or extreme emotional states that disrupt belief formation, larger errors are possible. Paradigmatically reliable processes like vision share these failings. Vision sometimes produces false beliefs under adverse conditions, or when we're looking at complex things. Still, it's so reliable as to be indispensible in ordinary life. Regarding phenomenal introspection as unreliable is about as radical as skepticism about the reliability of vision.

While contemporary psychologists reject introspection into one's motivations and other psychological causal processes as unreliable, phenomenal introspection fares better. Daniel Kahneman (2000), for example, writes that "experienced utility is best measured by moment-based methods that assess the experience of the present." Even those most skeptical about the reliability of phenomenal introspection, like Eric Schwitzgebel (2008), concede that we can reliably introspect whether we are in serious pain.[17] Then we should be able to introspectively determine what pain is like. So I assume the reliability of phenomenal introspection.

One can form a variety of beliefs using phenomenal introspection. For example, one can believe that one is having sound experiences of particular noises and visual experiences of different shades of color. When looking at a lemon and considering the phenomenal states that are yellow experiences, one can form some beliefs about their intrinsic features – for example, that they're bright experiences. And when

[17] Schwitzgebel offers counterexamples to the reliability of phenomenal introspection which usually show instead that false beliefs about our experiences can be formed by processes other than phenomenal introspection.

considering experiences of pleasure, one can make some judgments about their intrinsic features – for example, that they're good experiences. Just as one can look inward at one's experience of lemon yellow and recognize its brightness, one can look inward at one's experience of pleasure and recognize its goodness.[18] When I consider a situation of increasing pleasure, I can form the belief that things are better than they were before, just as I form the belief that there's more brightness in my visual field as lemon yellow replaces black. And when I suddenly experience pain, I can form the belief that things are worse in my experience than they were before.

Having pleasure consists in one's experience having a positive hedonic tone. Without descending into metaphor, it's hard to give a further account of what pleasure is like than to say that when one has it, one feels good. As Aaron Smuts (2011) writes in arguing that pleasure is a hedonic tone, "to 'feel good' is about as close to an experiential primitive as we get."[19] Fred Feldman (2004) sees pleasure as fundamentally an attitude rather than a hedonic tone. But as long as hedonic tones are real components of experience, phenomenal introspection will reveal pleasure's goodness. Opponents of the hedonic tone account of pleasure usually concede that hedonic tones exist, as Feldman seems to in discussing "sensory pleasures," which he thinks his view helps us understand. Even on his view of pleasure, phenomenal introspection can produce the belief that some hedonic tones are good while others are bad.

There are many different kinds of pleasant experiences. There are sensory pleasures, like the pleasure of tasting delicious food, receiving a massage, or resting your tired limbs in a soft bed after a hard day. There are the pleasures of seeing that our desires are satisfied, like the pleasure of winning a game, getting a promotion, or seeing a friend succeed. These experiences differ in many ways, just as the experiences of looking at lemons and the sky on a sunny day differ. It's easy to see the appeal of Feldman's view that pleasures "have just about nothing in common phenomenologically" (79). But just as our experiences in looking at lemons and the sky on a sunny day have brightness in common, pleasant experiences all have "a certain common quality – feeling good," as Roger Crisp (2006) argues (109). As the analogy with brightness suggests, hedonic tone is phenomenologically very thin, and usually mixed with a variety of other experiences.[20] Pleasure of any kind feels good, and displeasure of any kind feels bad. These feelings may or may not have bodily location or be combined with other sensory states like warmth or pressure. "Pleasure" and "displeasure" mean these thin phenomenal states of feeling good and feeling bad. As Joseph Mendola (2006) writes, "the pleasantness of physical pleasure is a

---

[18] Occasionally I doubt that pleasure is really good. Seeking evidence, I eat or drink something with a pleasant taste. When I consider my experience, I become convinced again.

[19] For similar views, see Locke (1690) and Hume (1888).

[20] For a similar analogy, but with the volume of sounds rather than the brightness of colors, see Kagan (1992).

kind of hedonic value, a single homogenous sensory property, differing merely in intensity as well as in extent and duration, which is yet a kind of goodness" (442).[21]

What if Feldman is right and hedonic states feel good in fundamentally different ways? Then phenomenal introspection suggests a pluralist variety of hedonism. Each fundamental flavor of pleasure will have a fundamentally different kind of goodness, as phenomenal introspection more accurate than mine will reveal. This isn't my view, but I suggest it to those convinced that hedonic tones are fundamentally heterogenous.

If phenomenal introspection reliably informs us that pleasure is good, how can anyone believe that their pleasures are bad? Other processes of moral belief formation are responsible for these beliefs. Someone who feels disgust or guilt about sex may not only regard sex as immoral but the pleasure it produces as bad. Even if phenomenal introspection on sexual pleasure disposes one to believe that it's good, stronger negative emotional responses to it may more strongly dispose one to believe that it's bad, following the emotional perception model suggested in the previous section. Explaining disagreement about pleasure's value in terms of other processes lets hedonists maintain that phenomenal introspection univocally supports pleasure's goodness. As long as negative judgments of pleasure come from unreliable processes instead of phenomenal introspection, the argument from disagreement eliminates them.

The parallel between yellow's brightness and pleasure's goodness demonstrates the objectivity of the value detected in phenomenal introspection. Just as anyone's yellow experiences objectively are bright experiences, anyone's pleasure objectively is a good experience.[22] While one's phenomenology is often called one's "subjective experience," facts about it are still objective. "Subjective" in "subjective experience" means "internal to the mind," not "ontologically dependent on attitudes towards it." My yellow experiences objectively have brightness. Anyone who thought my yellow experiences lacked brightness would be mistaken. Pleasure similarly is objectively good. It's true that anyone's pleasure is good. Anyone who denies this is mistaken. As Mendola writes, the value detected in phenomenal introspection is "a plausible candidate for objective value" (712).

Even though phenomenal introspection only tells me about my own phenomenal states, I can know that others' pleasure is good. Of course, I can't phenomenally introspect their pleasures, just as I can't phenomenally introspect pleasures that I'll experience next year. But if I consider my experiences of lemon yellow and ask what it would be like if others had the same experiences, I must think that they would be having bright experiences. Similarly, if in a pleasant moment, I consider

[21] Or from Mendola (1990), "Each moment of phenomenal experience presents itself as positively or negatively valenced to some degree. The valence is either (a) a lesser or greater degree of intrinsic value, as in the phenomenal component of happiness and pleasure; (b) a lesser or greater degree of intrinsic disvalue, as in the phenomenal component of unhappiness and pain; or (c) a null valence, which is neither intrinsic value nor disvalue. That valence of the experience of a moment is an objective phenomenal property." See also Tannsjo (2007).

[22] Here I use "goodness" and "value" interchangeably, as is customary.

what it's like for others to have exactly the experience I'm having, I must think that they're having good experiences. If they have exactly the same experiences I'm having, their experiences will have exactly the same intrinsic properties as mine. This is also how I know that if I have the same experience in the future, it'll have the same intrinsic properties. Even though the only pleasure I can introspect is mine now, I should believe that others' pleasures and my pleasures at other times are good, just as I should believe that yellow experienced by others and myself at other times is bright. My argument thus favors the kind of universal hedonism that supports utilitarianism, not egoistic hedonism.

Now I outline how the reliability of phenomenal introspection helps hedonism answer Joyce and Street's evolutionary debunking arguments. Joyce and Street argue that since evolution is selected for moral beliefs that promote survival and reproduction regardless of their truth, our moral beliefs are unlikely to be objectively true. We can respond by finding a psychological process that produces moral belief and is evolutionarily selected for reliability, as our sensory processes are. Phenomenal introspection is this process. Humans wouldn't be able to get around in the world and socially interact without knowing what their own experiences were like. Since belief in pleasure's goodness is a byproduct of phenomenal introspection, which is selected for reliability, it's reliably caused even if other moral beliefs aren't.[23] Belief in pleasure's goodness need not be specifically selected for truth, just as belief in astrophysics need not be selected for truth to avoid debunking arguments against astrophysics. All we need is a reliable underlying process. With astrophysics, we have sensory perception combined with empirical and mathematical reasoning; with morality, we have phenomenal introspection. If all other moral beliefs are debunked by their origins in processes not selected for reliability, an evolutionary debunking argument could do the same work for hedonists that the argument from disagreement.[24]

A full moral theory encompassing rightness and virtue can be built from the empirical results of phenomenal introspection along with conceptual analysis. Phenomenal introspection reveals pleasure to have a kind of goodness that makes states of affairs better in consequentialist moral theories. Conceptual analysis then connects the concept of goodness with the concept of a better state of affairs, and with other moral concepts like rightness and virtue. Even if conceptual analysis can't connect the moral and the nonmoral as a full normative ethical theory requires, it connects moral concepts to each other. The following propositions or something like them seem to be conceptual truths: states of affairs are pro tanto better insofar as they include more goodness, an action is pro tanto better insofar as it causally contributes to better states of affairs, and agents are pro tanto more virtuous insofar as they desire that better states of affairs obtain. These putative conceptual truths about pro tanto relations don't rule out strong forms of deontology. It's still

---

[23] This provides a targeted debunking argument of the sort considered in Kahane (2011). To see how hedonists are poised to use such arguments, see Lerner (n.d.).

[24] Street considers defenses of realism grounded in pleasure's goodness and pain's badness, but these arguments don't regard pleasure and displeasure as hedonic tones.

conceptually possible for obligations to trump good consequences in determining right action. Utilitarians building theories along these lines can treat deontology as a conceptually coherent position whose substantive claims lack any empirical evidence. The argument from disagreement provides empirical reasons to abandon belief in rightness that isn't reducible to pleasure's goodness. If the argument from disagreement forces us to abandon belief in all other moral facts, introspecting pleasure's goodness and following these conceptual pro tanto connections to other moral concepts will be the only way to develop a full moral theory.

Pleasure's Goodness Is Moral Value [6]

This section argues that the goodness of pleasure detected in phenomenal introspection is moral value, rather than the value of some other kins. "Good" also expresses many nonmoral value concepts. These include aesthetic, functional, and prudential value.[25] So one might object that phenomenal introspection reveals pleasure only to have some kind of nonmoral value, which hedonists mistake for moral value. It would be nice to offer a conceptual analysis of moral value which entails the moral value of pleasure, but Moore's objections to doing normative ethics by conceptual analysis rule out such a straightforward answer. So I sketch a metaethical picture, consisting of synthetic reductive naturalism and an experientialist analysis of moral concepts, that helps hedonists argue that phenomenal introspection reveals genuine moral value.

Hedonists can accept synthetic a posteriori identity relations between moral properties and natural properties, following Peter Railton (1986, 1989). Just as electrolyzing water suggests that water is H_2O, the argument from disagreement and phenomenal introspection together suggest that goodness is pleasure. In both cases, empirical evidence suggests property identities. While identity permits moral properties to be where phenomenal introspection could find them, it doesn't guarantee this. If X and Y are identical as a matter of synthetic a posteriori necessity, one can know that X is present without being in position to know that Y is present. The advantage of synthetic reductive naturalism is simply that it puts moral properties in the natural world where they might be empirically accessible. To empirically determine whether phenomenal introspection actually reveals moral properties to apply to pleasure, we should check by using introspection, as some philosophers do. In "The Appeal of Utilitarianism," Robert Shaver (2004) writes that "The goodness or badness of pleasures or pains is evident from the point of view of anyone capable of experience" (248). In *Normative Ethics,* Shelly Kagan (1998) similarly notes that "The value of pleasure and the disvalue of pain seem virtually self-evident to anyone experiencing them" (30). These philosophers seem to be detecting moral

[25] Many of these aren't plausible alternatives. Phenomenal introspection doesn't reveal that pleasure has functional value like a sharp knife, or that it has aesthetic value like *Macbeth*.

properties in experience, one of the many places where synthetic reductionism allows them to be.

Opponents of ethical hedonism might argue that these philosophers are misclassifying the goodness of pleasure, and that phenomenal introspection merely reveals pleasure to be good in some nonmoral way. To answer this objection, I present a conceptual analysis of moral terms and a theory of reference that together explain how phenomenal introspection can reveal pleasure's moral value.

Experientialism, which I defend in *Humean Nature,* is an analysis of moral concepts in terms of what moral feelings objectively represent. Feelings are the experienced effects of emotional dispositions, distinguished from the motivational and physiological effects. Moral feelings represent the world as having moral properties, much as color experience represents the world as having color properties. Experiencing guilt about something you did is feeling that it's wrong, much as experiencing blueness while looking at the sky is seeing it as blue. Here are some experientialist analyses of moral concepts:

GOOD	States of affairs objectively represented by accurate hope and delight
RIGHT	Actions objectively represented by accurate pride and approval
VIRTUE	Character traits objectively represented by accurate admiration
BAD	States of affairs objectively represented by accurate horror and sorrow
WRONG	Actions objectively represented by accurate guilt and indignation
VICE	Character traits objectively represented by accurate contempt and hatred

Experientialism doesn't connect moral concepts with motivation or reasons. This makes it fully externalist. It gives moral concepts more content than externalism often does by connecting them with what feelings represent. Saying that war is bad is saying that horror and sorrow objectively represent what it is like, morally speaking.

What does experientialism say about moral properties and how we represent them? It demands that moral judgments have objective content, requiring that any moral properties be objective. Apart from that, it doesn't settle the nature of moral properties. They could be nonnatural, natural, or nonexistent. It also leaves open how moral feelings might represent things. Perhaps we represent these properties via nonnatural reference magnets, or via rigid designation and a causal theory of reference. The answers to these questions are synthetic, so they aren't in the experientialist conceptual analysis.

My own answer to these questions is that moral feelings represent natural properties in virtue of shared phenomenal character, much as empathy represents how others feel. If moral feelings represent moral properties via this kind of empathic representation, the hedonic character of moral feelings lets them represent hedonic properties. Then the moral value of pleasure can be detected in introspection. A defense of this view would outlast my readers' patience. So the rest of this section just states the view, leaving a defense for another time.

Some empathic experiences represent the experiences of those we empathize with in virtue of shared phenomenal character. If my friend feels sad and I empathically feel sad as I see her, I represent her feelings accurately. Empathy lets me know how she feels. But if it she's actually happy and I feel sad for her, my experience misrepresents her feelings. Empathic accuracy is a matter of sharing the phenomenal character of her experience.

The moral feelings invoked by experientialism all have a hedonic character. Moral feelings representing positively valenced moral properties like goodness, rightness, and virtue are pleasant. These include hope, delight, pride, approval, and admiration. Moral feelings representing negatively valenced moral properties like badness, wrongness, and vice are unpleasant. These feelings include horror, sorrow, guilt, indignation, contempt, and hatred.

Applying an empathic account of representation to moral feelings suggests a unified hedonistic account of what all moral feelings represent. Positive moral feelings represent increasing hedonic tone and relations to it, while negative moral feelings represent decreasing hedonic tone and relations to it. Moral feelings are isohedonic to what they represent, sharing its hedonic phenomenal character. If you're horrified that war is breaking out, your horror is isohedonic with the suffering that will result. Both include displeasure. Someone who was delighted that war was breaking out wouldn't have feelings isohedonic with this suffering. This is why delight misrepresents the moral character of war. (If it's one of the rare wars that increases the total hedonic tone of reality, delight represents it accurately while horror is inaccurate. Wars that one should feel good about are unusual.) If you admire a leader who makes peace, your pleasant feeling of admiration has a rising hedonic tone matching the hedonic improvement that the leader desired, or perhaps matching the happy consequences that such peacemaking dispositions typically generate.[26] The hedonic match between admiration and such desired or typical consequences explains why peacemakers are admirable, or in other words, virtuous.

Experientialism and the empathic account of representation entail that we can introspectively detect the moral value of pleasure. Introspection reveals that pleasure objectively shares the pleasant character of hope and delight. According to the empathic account, hope and delight accurately represent things that share their pleasant character. And according to experientialism, things objectively represented by hope and delight have moral value. So introspection shows that pleasure has moral value.

Conclusion [7]

First I argued that we should give up all our moral beliefs, as widespread moral disagreement reveals the general unreliability of the processes producing them. Then I argued that phenomenal introspection reliably informs us of pleasure's

[26] These different accounts of how virtue relates to good states of affairs are suggested by Hurka (2001) and Driver (2001). Pluralism about how virtue relates to good states of affairs, encompassing both accounts, is plausible here.

goodness. This is a counterexample to premise 4 of the argument from disagreement. Premises 1–3 of that argument still stand, entailing 6: "We should retain only those moral beliefs formed through reliable processes." If phenomenal introspection is the only reliable process for discovering synthetic moral truths, our only moral beliefs should be in the goodness of pleasure and whatever it entails. We should accept a version of ethical hedonism on which we doubt rather than deny nonhedonic moral facts.

This gives hedonists a powerful response to the intuitively compelling counterexamples that lead many to reject their view and believe in a wider range of moral facts. Hedonists can respond that while it intuitively seems that many kinds of moral goodness, badness, rightness, wrongness, virtue, and vice aren't grounded in pleasure's goodness, beliefs in such moral facts are unacceptably likely to be mistaken. Rejecting hedonism is intuitive to us only because we're afflicted by unreliable processes of belief formation. To avoid moral error, we should form moral beliefs only through reliable processes. Then pleasure is the only thing we'll believe is good. The conceptual connections between moral properties of things, states of affairs, actions, and characters provide only reliable route to a full moral theory.

This argument for hedonism won't make hedonism feel right. The epistemic processes that have led so many humans into error make the putative counterexamples to hedonism seem convincing. So hedonism will feel wrong even to hedonists, just as optical illusions look the same even after we know the truth. It feels like this to face our strongest cognitive biases. We should expect to feel this way, given how much moral error has historically existed and how deeply entrenched it is in our moral feelings. Whatever the moral truth is, it would feel wrong to most humans. So it's likely that the moral truth would feel wrong to us.

Methodological advances have contributed to human knowledge in many areas of inquiry. Nothing short of a methodological revolution could take us from the widespread moral error that has afflicted humankind to knowledge of the moral truth. If my arguments are sound, the truth is that pleasure is the only thing we can know to be good. Relying solely on phenomenal introspection to reveal the nature of goodness is the revolution.[27]

[27] This chapter circulated unpublished for over a decade before Sanjit Chakraborty kindly included it in the present volume. It is discussed by Hookom (2011), de Lazari-Radek and Singer (2014), Wren-Lewis (2015), Bertrand (2016), Kammerer (2019), Sobel (2019), Andrew (2021), Schroeder (2021), Broi (2022), Crisp (2022), and Delon (2022). This version of the chapter remains close to the cited draft, revised mainly for Routledge house style and to add this footnote. Its main arguments are now appearing elsewhere. My (forthcoming-c) "The reliable route from nonmoral evidence to moral conclusions" presents the foundations of the epistemic argument. My (forthcoming-b) "Pleasure is goodness; morality is universal" explains how experientialism, which I now call "moral sense theory," favors hedonism. My (forthcoming-a) "Naturalistic arguments for ethical hedonism" summarizes both arguments. I thank many friends and teachers at Harvard, Texas, Michigan, and National University of Singapore (NUS) for helping me develop this defense of hedonism, which occurred to me as a college freshman in 1998 and led me to study philosophy.

References

Andrew, J. P. (2021). *Freedom, mattering, and the intrinsic badness of suffering* (Dissertation). University of Texas at Austin.

Arpaly, N. (2003). *Unprincipled virtue*. Oxford University Press.

Barber, A. (2013). Science's immunity to moral refutation. *Australasian Journal of Philosophy, 91*(4), 633–653.

Bertrand, O. (2016). Reducing suffering or adding years of life? *14th conference of the International Society for Utilitarian Studies.*

Brink, D. (1989). *Moral realism and the foundations of ethics* (pp. 197–209). Cambridge University Press.

Broi, A. (2022). Une défense de l'hédonisme axiologique. *Dialogue: Canadian Philosophical Review/Revue canadienne de philosophie, 61*(2), 325–346.

Comesaña, J. (2006). A well-founded solution to the generality problem. *Philosophical Studies, 129*, 27–47.

Conee, E., & Feldman, R. (1998). The generality problem for reliabilism. *Philosophical Studies, 89*, 1–29.

Crisp, R. (2006). *Reasons and the good* (pp. 107–109). Oxford University Press.

Crisp, R. (2022). Deep personal relationships and well-being: A response to Hooker. *Ratio, 35*(4), 301–309.

de Lazari-Radek, K., & Singer, P. (2014). *The point of view of the universe: Sidgwick and contemporary ethics*. Oxford University Press.

Decker, J., & Groll, D. (2013). On the (in)significance of moral disagreement for moral knowledge. In *Oxford studies in metaethics* (Vol. 8). Oxford University Press.

Delon, N. (2022). Strangers to ourselves: A Nietzschean challenge to the badness of suffering. *Inquiry*, 1–30.

Doris, J., & Plakias, A. (2008). How to argue about disagreement: Evaluative diversity and moral realism. In Sinnott-Armstrong (Ed.), *Moral psychology* (Vol. 2, pp. 303–331). MIT Press.

Driver, J. (2001). *Uneasy virtue*. Cambridge University Press.

Elga, A. (2007). Reflection and disagreement. *Noûs, 41*(3), 478–502.

Enoch, D. (2009). How is moral disagreement a problem for realism? *The Journal of Ethics, 13*, 15–50.

Feldman, F. (2004). *Pleasure and the good life*. Oxford University Press.

Foucault, M. (1977). *Discipline and punish: The birth of the prison.*, Sheridan-Smith (trans.). Vintage.

Goldman, A. (2000). What is justified belief? In Sosa & Kim (Eds.), *Epistemology: An anthology*. Blackwell.

Haidt, J. (2001). The emotional dog and the rational tail. *Psychological Review, 108*(4), 814–834.

Harman, G. (1988). Ethics and observation. In G. Sayre-McCord (Ed.), *Essays on moral realism*. Cornell University Press.

Homer. (1900). *The Odyssey*, Butler (trans.). AC Fifield.

Hookom, A. (2011). *But what kind of badness?: An inquiry into the ethical significance of pain* (Thesis). Georgia State University.

Huemer, M. (2005). *Ethical intuitionism*. Palgrave MacMillan.

Hume, D. (1888). In L. A. Selby-Bigge (Ed.), *Treatise of human nature*. Clarendon Press.

Hurka, T. (2001). *Virtue, vice and value*. Oxford University Press.

Jackson, F. (1998). *From metaphysics to ethics*. Oxford University Press.

Jost, J., Banaji, M., & Nosek, B. (2004). A decade of system justification theory: Accumulated evidence of conscious and unconscious bolstering of the status quo. *Political Psychology, 25*(5), 881–919.

Joyce, R. (2001). *The myth of morality*. Cambridge University Press.

Kagan, S. (1992). The limits of well-being. In E. F. Paul, F. D. Miller Jr., & J. Paul (Eds.), *The good life and the human good*. Cambridge University Press.

Kagan, S. (1998). *Normative ethics*. Westview Press.
Kahane, G. (2011). Evolutionary debunking arguments. *Nous, 45*(1), 103–125.
Kahneman, D. (2000). Experienced utility and objective happiness: A moment-based approach. In Kahneman & Tversky (Eds.), *Choices, values and frames*. Cambridge University Press.
Kammerer, F. (2019). The normative challenge for illusionist views of consciousness. *Ergo, an Open Access Journal of Philosophy, 6*, 891–924.
Kauppinen, A. (2013). A Humean theory of moral intuition. *Canadian Journal of Philosophy, 43*(3), 360–381.
Kelly, T. (2005). The epistemic significance of disagreement. In Hawthorne & Szabo (Eds.), *Oxford studies in epistemology* (Vol. 1). Oxford University Press.
Lerner, A. (n.d.). *Fine-tuning evolutionary debunking arguments*, unpublished.
Locke, J. (1690). *Essay concerning human understanding*. Bassett.
Mackie, J. (1977). *Ethics: Inventing right and wrong*. Penguin.
Mendola, J. (1990). Objective value and subjective states. *Philosophy and Phenomenological Research, 50*(4), 695–713.
Mendola, J. (2006). Intuitive hedonism. *Philosophical Studies, 128*, 441–477.
Merli, D. (2008). Expressivism and the limits of moral disagreement. *The Journal of Ethics, 12*, 25–55.
Moore, G. E. (1903). *Principia ethica*. Cambridge University Press.
Nietzsche, F. (1998). *On the genealogy of morality*., Clark and Swensen (trans.). Hackett.
Nozick, R. (1974). *Anarchy State and Utopia*. Basic Books.
Railton, P. (1986). Moral realism. *Philosophical Review, 95*(2), 163–207.
Railton, P. (1989). Naturalism and prescriptivity. *Social Philosophy and Policy, 7*(1), 151–174.
Sayre-McCord, G. (1988). Moral theory and explanatory impotence. In G. Sayre-McCord (Ed.), *Essays on moral realism*. Cornell University Press.
Schnall, S. (2008). Disgust as embodied moral judgment. *Personality and Social Psychology Bulletin, 34*(8), 1096–1109.
Schroeder, M. (2008). *Being for*. Oxford University Press.
Schroeder, M. (2021). A common subject for ethics. *Mind, 130*(517), 85–110.
Schwitzgebel, E. (2008). The unreliability of naive introspection. *Philosophical Review, 117*(2), 245–274.
Shaver, R. (2004). The appeal of utilitarianism. *Utilitas, 16*(3), 235–250.
Sinhababu, N. (2017). *Humean nature*. Oxford University Press.
Sinhababu, N. (forthcoming-a). Naturalistic arguments for ethical hedonism. In R. Y. Chappell, D. Meissner, & W. MacAskill (Eds.), *An introduction to utilitarianism*.
Sinhababu, N. (forthcoming-b). Pleasure is goodness; morality is universal. *Ethical Theory and Moral Practice*, 1–17.
Sinhababu, N. (forthcoming-c). The reliable route from nonmoral evidence to moral conclusions. *Erkenntnis*, 1–21.
Smuts, A. (2011). The feels good theory of pleasure. *Philosophical Studies, 155*, 2.
Sobel, D. (2019). The case for stance-dependent reasons. *Journal of Ethics and Social Philosophy, 15*, 146.
Stewart, T. (2005). The competing practices argument and self-defeat. *Episteme, 2*, 13–24.
Street, S. (2006). A Darwinian dilemma for realist theories of value. *Philosophical Studies, 127*(1), 109–166.
Tannsjo, T. (2007). Narrow hedonism. *Journal of Happiness Studies, 8*, 79–98.
Tappolet, C. (2016). *Emotions, values, and agency*. Oxford University Press.
Vavova, K. (2014). Moral disagreement and moral skepticism. *Philosophical Perspectives, 28*, 302–333.
Werner, P. (2016). Moral perception and the contents of experience. *Journal of Moral Philosophy, 13*(3), 294–317.
Wren-Lewis, S. (2015). Towards a widely acceptable framework for the study of personal well-being. In J. Søraker, P. Wong, J. Boer, J. De Rijt, & P. Brey (Eds.), *Well-being in contemporary society* (pp. 17–38). Springer.

Neil Sinhababu is Associate Professor of Philosophy at the National University of Singapore. He is author of Humean Nature (2017) and co-editor of Nietzsche and Morality (2007), both published by Oxford University Press. His articles have appeared in journals including Philosophical Review, Noûs, and the Journal of Philosophy. His research centers on metaethics and includes epistemology, metaphysics, philosophy of mind, and Nietzsche. As he enjoys discussing all these topics with other philosophers, he once gave invited lectures at over 100 universities in 1 year.

The Interpreting Mind and Meaning-Formation: A Relational Critique of Lyotard and Ricœur

Gerald Cipriani

Abstract Jean-François Lyotard in *Discourse, Figure* interprets the phenomenon of meaning-formation from the Lacanian standpoint of the unconscious, desire, and *jouissance*. There is, however, always the danger that the order of discourse be replaced by the order of *jouissance*, for such an order of interpretation comprehends meaning-formation as a discrete entity, that is, in non-relational terms. Similarly, in *The Conflict of Interpretations*, Paul Ricœur's descriptive hermeneutics of negotiation falls short of fully considering the intrinsically relational nature of meaning-formation in so far as his way of bringing different methods of interpretation into relation with each other (structural semiotics, Freudian psychoanalysis, and phenomenology) amounts to a comparison between discrete entities.

This chapter unfolds a relational critique of Lyotard's determinant mode of interpretation and Ricœur's hermeneutic methodology to pave the way for a configuration whereby each element involved (such as interpreting minds, values, and physical objects) should be understood as "dependently co-arisen", to use the established Buddhist phrasing. Each method of interpretation therefore contributes to a holistic configuration that makes up the stuff of "reality" through a dynamic of mutual balancing between complementary, differential entities. The issue is therefore not only reflexive and epistemological but also mesological and ethical.

Keywords Interpretation · Poststructuralism · Hermeneutics · Aesthetics · Relation · Co-dependence

> Each thing necessarily has some place from which it can be affirmed as thus and so, and some place from which it can be affirmed as acceptable. So no thing is not right, no thing is not acceptable. For whatever we may define as a beam as opposed to a pillar, as a leper as opposed the great beauty Xishi, or whatever might be [from some perspective] strange,

G. Cipriani (✉)
Graduate Institute of Philosophy, National Central University, Taoyuan City, Taiwan
e-mail: gcipriani@ncu.edu.tw

S. Chakraborty (ed.), *Human Minds and Cultures*,
https://doi.org/10.1007/978-3-031-56448-2_9

grotesque, uncanny, or deceptive, there is some course that opens them into one another, connecting them to form a oneness.[1] Zhuangzi

物固有所然 物固有所可。無物不然 無物不可。故爲是擧莛與楹 厲與西施 恢恑憰怪 道通爲 一。莊子

Prelude

What is more difficult than to understand the experience of meaning? Hermeneutics seeks to work out what is happening when understanding meaning in art, literature, or the scriptures, among others. But if understanding the experience of meaning simply amounts to establishing validity in interpretation according to pre-managed criteria, then we can justifiably qualify this kind of endeavour as "economy".[2]

When Jean-François Lyotard in *Discours, figure* (1971) interprets the phenomenon of meaning-formation from the Lacanian standpoint of the unconscious,[3] what he does is willingly (or unwillingly) elaborate an "economy of desire", a "symptomology of origins", or even a "methodology of the repressed". Clearly, what for Lyotard justifies such an interpretation is the *topos* of meaning-formation, that is, "discourse" as elaborated in Western culture. With cultural Modernism, time had come to understand meaning-formation as release or liberation from discursive order, triggering thus a "*jouissance*" of the kind experienced by the slave getting rid of the grip of the master. There is, however, always the danger that the order of discourse be replaced by the order of *jouissance* not least because a methodology of the repressed does not consistently attend to the specificity of its field such as with ideology but because such an order of interpretation comprehends meaning-formation as a discrete entity, that is, in non-relational terms.

In such a case, the *jouissance* becomes a phenomenon in its own right whose reading smacks of the kind of essentialism that tells us what meaning-formation is truly made of. In actual fact, meaning-formation comes into view at a particular time and location not only because the order of interpretation simply concurs with the phenomenon in its singular dimension but also because the latter becomes manifest through that wherein it relates and their correlative modes of interpretation. In the case of Lyotard's study of meaning-formation in the context of Modernism, the liberation of the "figure" from the grip of discourse, far from attesting to some specified authenticity of art (i.e. ecstatic release) calling for the relevant interpretation (i.e. psycho-critical analysis), is a manifestation whose nature owes to what it relates within the fabric of culture at a particular time and location. The place of the manifestation, as it were, is made of the institution, the canons of representation, the

[1] Zhuangzi, "Equalizing Assessments of Things," in *Zhuangzi: The Essential Writings*, trans. Brook Ziporyn (Indianapolis: Hackett Publishing Company, 2009), 13.

[2] Unless mentioned otherwise, translations are mine.

[3] Jean-François Lyotard, *Discours, figure* (Paris: Klincksieck, 1971).

tradition of metaphysics, semiotic structure, collective values, the artist's intention, and so on, in other words, the myriad of elements whose interconnected dimension together with their corresponding interpretations makes meaning-formation as release, liberation, or *jouissance* visible. From this angle, no interpretation can provide the truth of meaning-formation universally—of course not—nor merely in relation to its object of interpretation. Meaning owes its truth to the relational junction between various modes of interpretation and the field of its expression. Even Paul Ricœur's descriptive hermeneutics of negotiation falls short of fully considering this aspect in so far as his way of bringing different methods of interpretation into relation with each other amounts to a comparison between discrete entities albeit for the sake of showing their *raison d'être* and providing a better understanding in the midst of relational conflicts.[4]

For Ricœur, the point is certainly not to replace an order of interpretation with another allegedly more faithful one, and even less to establish a new one. When pondering on hermeneutic validity in his *Le conflit des interprétations* (1969), he suggests a dialectical confrontation between different modes of interpretation—structuralist semiotics, Freudian psychoanalysis, and phenomenology—in order to highlight their concrete relevance to a particular "problem" pertaining to meaning-formation. Instead of being determinant as in the case of Lyotard, the method is reflexive, in the sense it reveals the conditions of the modes of interpretation for a given, concrete situation. The method shows, for example, that structuralist semiotics cannot provide an account of meaning from the standpoint of an unconscious origin any more than Freudian psychoanalysis can allow for the eventful nature of meaning-formation in terms of novelty. The reflexivity of Ricœur's descriptive hermeneutics is assuredly a necessary step for anyone who wants to come as close as can be to understanding different modes of interpretations, that is, to become aware of their benefits and limitations. But the hermeneutics does not give enough weight to understanding meaning-formation in intrinsic relational terms, leading thus inexorably to what has characterised much of the tradition of Western thought, that is, the abstraction by the mind of phenomena and values.

To focus on meaning-formation in an enlightened manner, namely without blindly imposing particular modes of interpretation and therefore the subject's particular perspective, requires uncovering the relational determining reciprocities at work among the various formative elements that constitute the oneness of a reality in the shaping, be they semiotic, psychoanalytical, or phenomenal; political, historical, or cultural. In other words, each of these interpretive elements should be understood as "dependently co-arisen" (as well as conventionally denominated), to use the Buddhist established phrasing and in particular that of Nagarjuna who sought to provide a more adequate understanding of the fundamental of "reality".[5] This

[4] See Paul Ricœur, *Le conflit des interprétations. Essais d'herméneutique* (Paris: Le Seuil 1969).

[5] See, for example, Kottegoda S. Warnasuriya, "Dependent Origination and the Buddhist Theory of Relativity," *Hsi Lai Journal of Humanistic Buddhism* (西來人間佛教學報) 8 (2007):153–163; Jay L. Garfield, "Nagarjuna's Theory of Causality: Implications Sacred and Profane," *Philosophy East and West* 51, 4 (October 2001): 507–524.

chapter will endeavour to uncover the relational condition at stake by critically discussing Lyotard's determinant mode of interpretation and Ricœur's hermeneutic methodology when it comes to understanding meaning-formation.

The Order of *Jouissance*

Meaning-formation is a phenomenon that applies to all areas of human experience such as everyday life, science, or culture. It is, however, within a specific area of the cultural sphere at the crossroad between expression and communication that the question of desire is the most telling: the arts. As soon as expression and communication are involved, we cannot do without norms, rules, codifications, symbols, and so on; that is, what in Lyotard's interpretation amounts to "discourse". In this context, discourse is not a simple formal discussion of a theme in writing or speech. Discourse is an imposed—and therefore potentially repressive, order that inexorably originates the desire to break off the conventionally established relationship between representation and its designated reality. From this perspective, the phenomenon of meaning-formation understood as liberation from discourse is a breaking-off that Lyotard qualifies in terms of "force" or "energy" of various intensities and modes and whose psychological correlative amounts to "*jouissance*", "suffering", or "pathos".

Thus, conceived meaning-formation as epitomised in Modernism is therefore no order of discourse; it does not amount to any order of representation or narration demanding recognition to do it justice. Meaning-formation is here "figural", as Lyotard calls it. Hence, through its unrecognisable nature, the figural deconstructs the order of discourse, as well as the familiar and that with which we feel at home such as the correct or the beautiful (in the narrow sense of the word).[6] The figure-meaning, unlike the represented or narrated figure, *becomes* manifest as we unlearn through it the order of discourse. But should this suggest that the figural is an abstraction or even that it is deprived of sense? Far from it. The figural *means* by liberating itself from what is felt to be a repressing order. If we transpose André Malraux' conception of creative style into Lyotard's psycho-critical context, we may qualify the figural as "coherent deformation".[7] The figural *makes* sense by deviating from the order of representation and narration before becoming itself

---

[6] "[*Le figural*] *ne déconstruit pas seulement le discours, mais la figure en tant qu'image reconnaissable ou forme bonne.*" Jean-François Lyotard, *Discours, figure* (Paris: Klincksieck, 1971), 328.

[7] The expression "*déformation cohérente*" comes from Malraux in *La Création artistique* (Genève: Skira, 1948), 152. Art proceeds to a proper metapmorphosis of existing styles and by doing so creates a new style. Coherent deformation is therefore valid means of creation ("*un moyen légitime de création*," 180). He gives the example of modernist painter Édouard Manet (1832–1883): "*Manet nous fait découvrir le bouquet et le vase curieusement cerné de l'*Infante Marguerite-Thérèse [...]. *La vie des grands styles brasse inlassablement le passé, parce que chacun d'entre eux le laisse "suspendu*"." Ibid., 212.

canonised into some recognisable style. Or, within the same paradigm, the figural is for Gilles Deleuze "non-sense" as it cannot be pinned down by the established logic of figuration. The figural, however, expresses paradoxes of the kind encountered, for instance, in Lewis Carrol and Antonin Artaud. In this respect, the figural is not to be identified as "absence of sense".[8] Meaning-formation thus understood as figurality possesses therefore a certain "opacity" that, to borrow the language of structural linguistics, contrasts with the transparency of the signifier component of the sign whose signified side points in the direction of a designated reality in the case of representation or to other signified units as in narration.

As such, particular works of art and historical periods can be identified as offering various degrees of figurality as well as contexts for desire to be liberated from the order of discourse. In his "*Veduta sur un fragment de l'histoire du désir*" Lyotard identifies periods of paradigmatic "vehicles" of desire spanning Western cultural history from the Renaissance to the second half of the nineteenth century, that is, before the start of Modernism.[9] In the visual arts, the advent and reign of Albertian rules of perspective with their laws and conventions typify the incarnation of the order of discourse at work. By contrast, Modernism is interpreted as the period during which the energy of desire that had been previously constrained by the order of representation and narration is released. To put it more crudely, Modernism is the period that puts an end to the "formations of desire" to reach for the real through liberation from the grip of discourse. There are, however, exceptions to confirm the rule, as Lyotard identified. Hans Holbein the Younger's *The Ambassadors* (1533) is one of them, and particular features in this painting make it a forerunner of Modernism.

Drawing partly from Jacques Lacan's interpretation of the painting in his *Les quatre concepts de la psychanalyse*, the "formations of desire" are here conveyed through the representation of the ambassadors Jean de Dinteville and Georges de Selve and their surrounding objects, while the anamorphosis across the lower part of the painting embodies the force or energy that frees the figure-meaning from the conventional Albertian order of representation allegedly at the heart of the question of desire.[10]

The figural has thus a proper disrupting "density" or "opacity" that, in Lyotard's words,

---

[8] See Gilles Deleuze, *Logique du sens* (Paris: Les Éditions de Minuit, 1969). What Deleuze calls "non-sense" certainly does not *mean* anything but at the same time it is no "absence of meaning", for non-sense proceeds to a "giveness of meaning". "*Le non-sens est à la fois ce qui n'a pas de sens, mais qui, comme tel, s'oppose à l'absence de sens en opérant la donation de sens. Et c'est ce qu'il faut entendre par non-sens.*" Ibid., 89.

[9] See Lyotard in "*Veduta sur un fragment de l'histoire du désir*", in *Discours, figure*, 161–208.

[10] See Jacques Lacan, "*L'anamorphose*," in *Les quatre concepts fondamentaux de la psychanalyse—livre XI* (Paris: Éditions du Seuil, 1973 [1963–64]), 75–84. Lacan expounds his well-known interpretive "analysis" of Holbein's painting whereby the anamorphosis incarnates the annihilation of the "subject" (a form of "castration"), a "lack" that Holbein makes visible by inviting viewing to decentre itself in order to identify what is painted—indeed a *jouissance* at the core of "vision".

> [...] deconstructs not only discourse but the figure, inasmuch as the figure is a recognizable image or a regular form. And underneath the figural: difference. Not just the trace, not just presence-absence, period, indifferently discourse or figure, but the primary process, the principle of disorder, the incitement to *jouissance*. Not some kind of interval separating two terms that belong to the same order, but an utter disruption of the equilibrium between order and non-order.[11]

Lyotard's psycho-critical interpretation that rests on the ability to identify a "principle of disorder", of which *jouissance* would be the symptom, obviously moves away from phenomenology and its focus on an intentionality based on the interdependency between the eye and what is seen. Indeed, phenomenology falls short of providing an origin-seeking explanation of meaning-formation—one rooted in the unconscious. Paul Cézanne's painting of the Montagne Sainte-Victoire (1904–1906), for instance, depicts an "event" within the actual field of vision instead of a visual "object". The *montagne* comes into sight through Cézanne's vision and rendering against the background of the regular order of representation and as *jouissance* and liberation from "discourse". This, phenomenology cannot grasp it.[12]

Be that as it may, causality had never been the explanatory component of experience that phenomenology attempted to work out, whether in its Husserlian methodological version, Merleau-Pontian pre-theoretical chiasmatic descriptions, or even more so in Maldineian accounts of sensed an-intentionality[13]—an idea on which Lyotard should have focused profusely if he was to truly comprehend the *montagne* as "event in the visual field" in contrast with an object seen by a seeing subject.

Does Lyotard's criticism of phenomenology transform the question of desire into a plea for an order of *jouissance*? (And the same could be asked for its political correlative, i.e. an order of liberation from discourse). Put differently, is his conception of the question of desire an abstraction overlooking the *milieu* made of dependently co-arisen values to which correspond specific methods of interpretation and from which his psycho-critical explanation stems?

Lyotard's analysis of the nature of meaning-formation certainly finds timely and topical niches that justify the method. In *The Ambassadors*, meaning-formation understood as *jouissance* breaks off the order of representation that surrounds the anamorphosis in the foreground by calling for a decentred vantage point to be able

---

[11] Jean-François Lyotard, *Discourse, Figure*, translated by Antony Hudek and Mary Lydon (Minneapolis: University of Minnesota Press, 2011), 328. "[...] *le figural, qui ne déconstruit pas seulement le discours, mais la figure en tant qu'image reconnaissable ou forme bonne. Et sous le figural, la différence, non la trace tout court, la présence-absence tout court, indifféremment discours ou figure, mais le processus primaire, le principe de désordre, la poussée à la jouissance; non un intervalle quelconque séparant deux termes dans le même ordre, mais une rupture absolue d'équilibre entre un ordre et un non-ordre.*" Lyotard, *Discours, figure*, 328.

[12] "*Que la Montagne Sainte Victoire cesse d'être un object de vue pour devenir un évènement dans le champs visuel, c'est cela que Cézanne désire, c'est cela que le phénoménologue espère comprendre, et que je crois qu'il ne peut pas comprendre.*" Ibid., 21.

[13] E.g. see Maurice Merleau-Ponty, "*Le doute de Cézanne*" (1945) and *L'Œil et l'esprit* (1964); Henri Maldiney *Regard, parole, espace* (1973), *Art et existence* (1985), and *L'Art, l'éclair de l'être* (1993).

to identify the skull as skull. In the case of Cézanne's *Montagne Sainte-Victoire*, what meaning-formation deconstructs is the pre-modernist worldview of meaning-as-representation. In both cases, the explanation of meaning-formation rests on the principle of fulfilled desire to be liberated from some constraining order of representation.

Such a mode of interpretation is certainly justified when topical and timely, but in all cases, meaning-formation (i.e. the phenomenon) and its place (i.e. the political, historical, and cultural contexts of both percipient and what is perceived) should be understood in relational terms, that is, as mutually determining so that representation and *jouissance* do not become the discrete ingredients of abstract orders of comprehension, in other words, ideologies. Representation should be seen as one of the necessary conditions of *jouissance* and vice versa, so that the question of desire does not become an order of *jouissance*, that is, when truth in artistic meaning is to be retrieved in an invisible origin regardless of the place of its formation, and that representation as a genuine component of the work of art does not mutate into a form of order of perception of reality regulated by established canons. In contrast, this relational conception of modes of interpretation and values call for a conception of holistic reality made of entities whose existence depends on that to which they relate. One can therefore talk about a harmonious configuration made of interpretations and values in constant process of transformation precisely because of its relational nature. And what justifies such or such mode of interpretation and valuation should be the simple urge to re-establish the relational harmony between the related entities, that is, perceived things as interpreted and valued.

The political dimension of art is another telling example. Meaning-formation as release from desire has an opacity whose justification is for Lyotard political. But for such an opacity to remain ethically meaningful, interpretation must acknowledge the relational nature of its occurrence, just as representation must be seen in the light of its breaking-off, for there is no representation in a representational world. This, Lyotard does not seem to acknowledge, at least not explicitly, and the way he validates his psycho-critical interpretation against Hegel's aesthetics is a case in point.

Clearly, if there is a philosophy that epitomises the ideology of representation, it is Hegel's. Who else has idealised more than him the signified side of signs to which artistic configurations must be devoted? When this devotion is unfulfilled, he names it the "symbolic period". With the "classical period" the devotion is at its best. And when art begins to lose its ability to be devoted to the "Idea" by finding the means to represent it, what Hegel names the "romantic period", then he foresees its death.[14] In other words, the standpoint from which Hegel unfolds his grand art historical narrative is representation. Art that becomes aware of itself or art that brings forth the sensible can only smack of death, but, perhaps, the greatest death of all is art as

[14] See G.W.F. Hegel's *Lectures on Aesthetics* [*Vorlesungen über die Ästhetik*], trans. by T.M. Knox as *Hegel's Aesthetics, Lectures on Fine Art* (Oxford: Clarendon Press, 1975). For explanations of the way he conceives the relationships between art and 'spirit' see Charles Taylor (1975) in *Hegel*, Chap. XVII, 465–479.

jouissance. Meaning-formation as release from desire turns Hegel's order of representation upside down. Meaning emerges not in the direction of what it should be but from where it never was.

For Lyotard, as we may expect, Hegel's overlooking of the question of desire is what leads him to conceive of art in its undervaluing opposition to philosophical and theological discourse. In other words, to redeem the true nature of art can only be experienced through the *jouissance* of being liberated from the order of discourse—a figurality whose opacity translates into the fusion between meaning and the sensible. To understand the true nature of art, one has to swap the order of representation for the order of *jouissance*. Thus conceived, meaning-formation takes place at the very moment when the desire-inducing distance between the representation and the represented real breaks off. No wonder, then, why for Lyotard neither phenomenology nor structural semiotics can grasp the alleged true nature of the event of meaning. Perceiving meaning-formation is bound to involve an element of intentionality, but Lyotard's argument is that any phenomenological focus on intentionality can only overlook the true nature of artistic meaning understood as liberation from a repressive order that imposes on art a relationship to some origin outside of itself, that is, the order of representation. The same applies to structural semiotics whose method does not allow for the event of meaning to be conceived as "relinquishment" (*dessaisissement*) from the order of discourse. In fact, any method of interpretation that claims to be closer to the truth—any truth, whether psycho-critical, phenomenological, or semiological—ignores its relational nature and as such not only becomes an "order" but also can only fail to find the grounds to justify itself, namely as endeavour to contribute to the balance of the underlying oneness of reality. Such an overlooking is precisely at the origin of conflicts of interpretations, to use Ricœur's postulation. Different methods condition the very truth they seek to retrieve or establish and as such create issues of incommensurability. Yet, we have to ask ourselves whether simply confronting such methods with the aim of uncovering their conditioning nature does not ultimately betray their intrinsic relational nature.

An order of interpretation whose centricity is desire, or rather the release from a condition created by an order at the source of the formations of desire, is in actual fact what Ricœur questions, this time with reference to Freudian psychoanalysis.[15] The point for Ricœur is not to swap between orders of interpretation, but rather to confront them with each other with the aim of revealing what they can do and, indeed, their limitations. Freud's "realism of origins" that reads the taking-place of meaning as "a return of the repressed" rests on the premise of the existence of a past traumatic event.[16] The method therefore consists of recovering in the figure-meaning a "manifestation" of some kind that belongs to the past, and more specifically to our childhood. This method of interpretation allegedly allows us to reach a truth of

[15] Ricœur, "*Herméneutique et psychanalyse*", in *Le Conflit des interprétations,* 173–337; *De l'interprétation. Essai sur Freud* (Paris: Éditions du Seuil, 1965).

[16] Le "*retour du refoulé*", Ricœur, *Le Conflit des interprétations*, 211.

which the figure-meaning is a symptom. As we know, Freud applied this origin-seeking method to Leonardo's paintings such as *Mona Lisa* (c. 1503–1519), *Saint John the Baptist* (c. 1513–1516), and *Virgin and Child with Sainte Anne* (c. 1503–1519) as well as Michelangelo's sculpture *Moses* (c. 1513–1515), or tragedies such as Sophocle's *Oedipus Rex* (performed between c. 430 and 426) and Shakespeare's *Hamlet* (c. 1599–1601).[17]

When for Lyotard understanding the nature of the figural demand that we see the order of representation and by extension "discourse" as the traumatic place at the origin of the liberation from such an order, Ricœur diagnoses a flaw in Freud's interpretation simply because the artistic symbol "does not [only] repeat our childhood, but it [also] explores our adult life".[18] In other words, there is no reason why the figure-meaning, or the taking-place of meaning, should be understood through a dynamic of "regression" rather than "progression"; there is no reason why what Ricœur calls the "symptomatology of dreams and neuroses" should take control of "the domain of creativity in consciousness".[19] Ricœur asks, Is meaning "[i]n the return of the repressed or in the rectification of the old by the new?"[20] Correlatively, we may ask whether methods of interpretation should be archaeological or renewing. The question here is of course about validity in interpretation, and the issue for Ricœur is to know how to "apply a specified philosophical method to a specified problem, that of the constitution of the symbol".[21] Indeed, how can we allow for the possibility of "novelty of value" if meaning owes its formation to a past condition to which it somehow still belongs?[22]

By moving beyond Lacan's structuralism that reads the unconscious as language and by introducing the concept of figural as event of figure-meaning, Lyotard partly overcomes the problem when he locates creativity in the liberation from the grip generated by an order of discourse and representation that developed in Western culture from the Renaissance on. Even so, if Lyotard's conception of meaning-formation as release from the repressive in the form of *jouissance* is certainly not an "archaeology of the subject",[23] it amounts to an authentication according to a certain pre-defined order that establishes and recognises truth-function within power

[17] See Sigmund Freud, *The Interpretation of Dreams* [*Die Traumdeutung*] (1899); *Leonardo da Vinci A Psychosexual Study of an Infantile Reminiscence* [*Eine Kindheitserinnerung des Leonardo da Vinci*] (1910); *A General Introduction to Psychoanalysis* [*Vorlesungen zur Einführung in die Psychoanalyse*] (1915–17).

[18] Ricœur, *Le Conflit des interprétations*, 235.

[19] Ibid., 236.

[20] Ibid., 244. Although Ricœur was here specifically referring to the meaning of religion ("*le sens de la religion*") the question also applies to psychoanalytic interpretation of art.

[21] Ibid., 289.

[22] Ibid., 240.

[23] Ibid., 52. "[...] *la psychanalyse a son fondement dans une archéologie du sujet, la phénoménologie de l'esprit dans une téléologie, la phénoménologie de la religion dans une eschatologie.*" "[...] psychoanalysis is founded upon an archeology of the subject, the phenomenology of spirit upon a teleology, and the phenomenology of religion upon an eschatology".

relationships. After all, is not psychoanalysis ultimately "the history of desire in its great debate with authority", in Ricœur's formulation?[24] Psychoanalysis seeks to decipher which desire stems from which exercise of authority, or which repressed desire comes from our childhood. Desire ends up being portrayed as precondition for the possibility of the real world. And the power of reification is said not to come from the conscious mind but from the unconscious. Understanding creativity becomes, as Ricœur reads it through the lens of psychoanalysis, a matter of establishing "relations between the effect of pleasure and the technique associated with the work of art".[25] There is thus no reason to believe that the analogy established between the perceived configuration and the unconscious enables us to understand more authentically the "enigma" of meaning-formation. What this particular form of psychoanalytical method of interpretation does though is to produce what Ricœur calls "an envisioned absence [*absence visée*] that, far from dispelling it, redoubles the initial enigma".[26] On this point, the limitations of such a hermeneutics of suspicion are exposed when contrasting the outcomes of the method with what is left out by ignoring other methods of interpretation that are relationally coexisting.

Relational Understanding

There is conflict when methods of interpretation become orders competing against each other for universal validity—not of the interpreted configuration but of these very methods as they seek confirmation of such a validity in the said configuration. Orders of interpretation are therefore no more than exercises of abstraction that justify themselves by means of the particular that becomes for that reason confirmative or even illustrative. Yet, confronting such methods to reveal their conditioning nature for the good cause of applying them to configurations in a fully informed manner does not remove what characterises any order of interpretation, that is, abstraction. The psycho-critical interpretation of meaning-formation becomes an abstracted order as soon as it loses sight of its relational condition. There is no such a thing as meaning as a form of release from whatever order if not in relation to this very order, be it discursive, phenomenal, semiological, political, historical, and so on—in other words, to what constitutes its *place* of interpretation. Each

---

[24] "[…] *l'histoire du désir dans son grand débat avec l'Autorité.*" Ricœur, *De l'interprétation. Essai sur Freud*, 225.

[25] Ricœur, *Le Conflit des interprétations*, 256. "[…] *rapports entre l'effet du plaisir et la technique de l'œuvre.*"

[26] Ibid., 262. "[…] *une absence visée qui, loin de la dissiper, redouble l'énigme initiale.*" Willis Domingo translates "*visée*" by "envisioned", but the French term also evokes the directedness of the mind and the intentional act of consciousness (intentionality). See *The Conflict of Interpretations: Essays in Hermeneutics*, trans. Willis Domingo et al., edited by Don Ihde, (Evanston: Northwestern University Press, 1974), 206.

interpretation—each entity involved in the process of interpreting—is factually "dependant arising" (*pratityasamutpada*) as illustrated by the following key Buddhist principle:

This is, because that is.
This is not, because that is not.
This ceases to be, because that ceases to be.[27]

Any drive to abstract the psycho-critical from its web of interconnectedness in order to point to the alleged authentic meaning of such or such a configuration or cultural practice is to develop inevitably a language of authenticity—in the case of Lyotard, most often against his will. True, unlike with psycho-critical interpretation neither structural semiotics nor phenomenology is able to grasp a conception of meaning as opacity or in retrospect of what is actually perceived. Structural semiotics cannot grasp the event-like opacity of meaning-formation that brings together signifier, signified, and designated, and phenomenology cannot see the invisible origin of meaning-formation embedded in some desire to be freed from whatever repressing order. Neither can Hegel's aesthetics uncover that hidden origin or phenomenal dimension. Still, should meaning-formation understood through the lens of psycho-critical interpretation be for that matter ascribed the status of "function of truth" (*fonction de vérité*), to use Lyotard's expression when qualifying the figural?[28] Or should meaning-formation rather be understood in more holistic relational terms?

Admittedly, Lyotard implicitly suggests a relational element in the figural, for it is "inhabited by the formations of desire".[29] In other words, by deconstructing or de-constituting the forms of desire generated by the orders of "figuring" and their correlative modes of interpretation meaning-formation expresses a *relation*; meaning is released from some condition retrospectively seen as subduing grip *through* the subversion of orders of figuring and interpretation, be they based on criteria of representation or appearing, be they based on political or historical criteria. This relational dimension, however, should be acknowledged as the very condition for the possibility of meaning and interpretation. Meaning as *jouissance* from the fulfilled desire of liberation generated by established modes of representation and interpretation can only take place *in relation* to established orders such as discourse, the phenomenal, the political, or the historical among many others. It would therefore be more pertinent to say that the "function of truth" rests in the relation, not in the form of the relation. Qualifying meaning-formation in terms "function of truth" from the perspective of desire and release amounts to establishing another order, another *logos*; that of *jouissance*. Put differently and within the context of Modernism, what justifies a reading of meaning-formation in terms of release and *jouissance* that does not amount to ideology is the relational circumstance of the

---

27 Thich Nhat Hanh, *The Heart of the Buddha's Teaching* (New York: Three River Press, 1999), 221–222.

28 Lyotard, *Discours, figure*, 281.

29 Ibid.

event as interpreted within the whole of the reality made of entities in constant need of balancing and rebalancing. Ignoring this relationality can only lead to abstraction and therefore to the order of ideology.

The point here is of course not to reject altogether Lyotard's critical version of Lacanian analysis or Freudian psychoanalysis as genuine methods of interpretation when timely and apropos. But besides the validity of discrete modes of interpretation and value-ascribing per se, the issue is also relational. As previously mentioned, Ricœur suggests a reflexive hermeneutics by means of dialectical confrontation between interpretive renewal and archaeology, progression and regression, and discovery and retrieval. The methodological confrontation is that between interpretation that does justice to the novelty of meaning and interpretation that retrieves meaning through the excavation of the unconscious. Only by proceeding to such a comparison can we indeed become aware of the relevance and limitations of methods of interpretation such as psychoanalysis. As Ricœur puts it in *The Conflict of Interpretations*,

> This invitation to investigate psychoanalysis itself by confronting it with other points of view which seem to be diametrically opposed allows us to glimpse the real meaning of its limits. [...] Psychoanalysis is limited by the very thing which justifies it, namely, its decision to know, in cultural phenomena, only that which falls under an economics [*une économique*] of desire and resistances.[30]

The methodology, in the sense of reflection on methods, involves therefore a variety of methods other than psychoanalysis precisely for the latter's *raison d'être* to come into view. The principle applies to all methods of interpretation. In this light, Ricœur claims, the symbol should be investigated by other methods whether "phenomenological, Hegelian, or even theological". And in the end, "it will be necessary to discover in the semantic structure of the symbol itself the *raison d'être* of these other approaches and of their relationship to psychoanalysis".[31]

This description of methods of interpretation, or what Ricœur calls a "second reading" with the example of psychoanalysis, is certainly vital to shield ourselves against the threat of ideologies, that is, the orders of interpretation at work. For example, we may see in Lyotard's psycho-critical interpretation of *The Ambassadors* or Western Modernism a felicitous choice of approach that shares analogies with the particular problem under scrutiny or the object of perception. But unless the method is exposed to other points of view for its *raison d'être* to be disclosed, it may easily smack of an attempt to see in the chosen particular an illustration, or even confirmation of what is prejudged in the interpretive order of *jouissance*. Yet, Ricœur's

[30] Ricoeur, *The Conflict of interpretation*, 207. "*Cette invitation à approfondir la psychanalyse elle-même, par le moyen de son affrontement avec d'autres points de vue qui paraissent diamétralement opposés, laisse entrevoir le sens véritable des limites de la psychanalyse.* [...] *La psychanalyse est limitée par cela même qui la justifie, à savoir sa décision de ne connaître dans les phénomènes de culture que ce qui tombe sous une économique du désir et des résistances.*" Ricœur, *Le Conflit des interprétations*, 264.

[31] "[...] *il faudra bien découvrir, dans la structure sémantique du symbole lui-même, la raison d'être de ces autres approches et de leur relation à la psychanalyse.*" Ibid.

endeavour also contains within itself some limitations. The *raison d'être* of the relationship between different methods of interpretation is precisely to enact the means by which they contribute to a holistic configuration that makes up the stuff of reality through a dynamic of mutual balancing between complementary, differential entities (whether in terms of interpreting minds, values, physical objects, and so on). The issue is therefore not only reflexive and epistemological but also mesological and ethical.

The truth of meaning-formation does not rest in any discrete conception of whatever method of interpretation, be it semiological, historical, political, or elaborated from the perspective of liberation or *jouissance*; neither in the object nor in the subject, and not even in the correlation or fusion between *noesis* and *noema*, that is, in the intentionality at work in a given space and at a particular time. The truth at stake lies in the *relation*. Let us take again the example of Modernism in the West. This cultural movement can certainly be seen as a breaking-off from tradition, the canons of representation and conventions by way of formalistic experimentation as much as, following Lyotard, a liberation from the order of discourse, or an autonomous development against particular socio-economic systems as theorised by Theodor Adorno and Peter Bürger.[32] The interpretation can be historical in so far as the movement took place at a specific time rendering thus that from which it breaks off historical, that is, Modernity, whether or not we consider it with Jürgen Habermas to be an "unfinished project".[33] But the lens through which we look at Modernism can also be, as we saw, psychoanalytical if we use the interpretative paradigm of trauma (originated by institutional authority), desire for liberation (formed through the orders of discourse), *jouissance* (the pleasure of freedom), and symptom (of the release from a repressing past). The interpretation can equally be political; Modernism is a reaction against not only the cultural forms of the advanced capitalism of the time but also its underlying economic forces seen to be shaping the course of culture. Besides, Modernism has been a theatre of experimentations of all kinds driven by the will for culture and art to become aware of themselves and show their ability to express their own "spirit". The movement thus equally calls for formalistic, semiological, and structuralist interpretation for the configurational innovations to be brought to light. Then, Modernist configurations are also events and phenomena as they take place, appear, and are perceived as such, here and now, and of which only descriptions of the performative or perceptual experiences can bring to mind the particularity. These are only examples, and the list is endless as long as there are interpretations with a *raison d'être*. But it would be unthinkable to conceive of each interpretation or, similarly, each interpreted entity as a discrete operation or fact. There is no such a thing as release without repression or authority, no liberation without form, no *jouissance* without its manifestation, no desire without

[32] See Theodor W. Adorno, *Aesthetic Theory* [*Ästhetische Theorie*], 1970; Peter Bürger, *Theory of the Avant-Garde* [*Theorie der Avantgarde*], 1974.

[33] See Jürgen Habermas, "Modernity Versus Postmodernity," trans. Seyla Ben-Habib, *New German Critique* 22 (Winter 1981): 3–14. ["*Die Moderne—ein unvollendetes Projekt*," in *Kleine Politische Schriften* (Frankfurt a.M.: Suhrkamp, 1981), 444–464].

fulfilment, and so on—just as their interpretative correlatives cannot operate without being emplaced within a constellation of related perspectives and values.

No interpretation, even when considered as mind-action, stems out of itself. Interpretations operate within a whole made of codependent entities that constitute the world in all their relational dynamics and in the direction of a yet-to-be-perceived world. Interpretations as such are therefore also codependant. Just as the world as we know it cannot physically prevail with a single, overwhelming entity implying thus *a certain degree* of harmony between its constituents, so do interpretations. And the task of confronting them should be directed towards disclosing their level of co-dependence with the aim of ensuring that none of them is believed to have the ability to universally and sempiternally prevail over the others. This relational conception adds another dimension to Ricœur's appeal for a reflexive methodology that would describe what kind of method can do what sort of job for what particular problem: a realisation that meaning-as-interpreted, whatever this may be, takes shape through its codependant relationships with the multiple entities that make up the oneness within which it finds itself. The event of meaning as release and *jouissance* and its correlative psycho-critical interpretation would not transpire if not understood through the stuff of form, structure, or history, through its political, deconstructive, or phenomenal complexions, among others and ad infinitum. In other words, a reflexive methodology must be complemented by a relational understanding of meaning-as-interpreted for any hope of establishing validity in interpretation, should this cease to be looking like a dream. Each focus on one co-arisen dependency rather than another should be justified as a tacit attempt to re-establish harmony among the constellation of inter-related truths. Moreover, to favour an interpretation over another can therefore only be momentary and justified when seeking to contribute to the harmony of a constantly mutating configuration of co-dependences. This is how the mind can act for the good sake of the reality of which it is a part.

Acknowledgement This publication is supported by the National Science and Technology Council of Taiwan. Project Reference: NSTC 111-2811-H-008-012 / NSTC 110-2410-H-008-048-MY3.

References

Adorno, T. W. (1984). *Aesthetic theory* (R. Hullot-Kentor, Trans.). Continuum.

Bürger, P. (2004). *Theory of the Avant-Garde* (M. Shaw, Trans.). University of Minnesota Press.

Deleuze, G. (1969). *Logique du sens*. Les Éditions de Minuit.

Freud, S. (1916). *Leonardo da Vinci a psychosexual study of an infantile reminiscence* (A. A. Brill, Trans.). Moffat, Yard & Company.

Freud, S. (1920). *A general introduction to psychoanalysis* (G. Stanley Hall, Trans.). Horace Liveright Publisher.

Freud, S. (1999). *The interpretation of dreams* (J. Crick, Trans.). Oxford University Press.

Garfield, J. L. (2001). Nagarjuna's theory of causality: Implications sacred and profane. *Philosophy East and West, 51*(4), 507–524.

Habermas, J., & Ben-Habib, S. (1981). Modernity versus postmodernity. *New German Critique, 22*, 3–14.
Hegel, G. W. F. 1975. *Hegel's Aesthetics, lectures on fine art* (T. M. Knox, Trans.). Clarendon Press.
Lacan, J. (1973). *Les quatre concepts fondamentaux de la psychanalyse—livre XI*. Éditions du Seuil.
Lyotard, J.-F. (1971). *Discours, figure*. Klincksieck.
Lyotard, Jean-François. (2011). *Discourse, figure* (A. Hudek & M. Lydon, Trans.). University of Minnesota Press.
Maldiney, H. (1973). *Regard, parole, espace*. Éditions L'Âge d'Homme.
Maldiney, H. (1985). *Art et existence*. Klincksieck.
Maldiney, H. (1993). *L'Art, l'éclair de l'être*. Les éditions du cerf.
Malraux, A. (1948). *La Création artistique*. Skira.
Merleau-Ponty, M. (1949). Le doute de Cézanne. In *Sens et non-sens* (pp. 15–49). Les Éditions Nagel.
Merleau-Ponty, M. (1964). *L'Œil et l'esprit*. Gallimard.
Ricœur, P. (1965). *De l'interprétation. Essai sur Freud*. Éditions du Seuil.
Ricœur, P. (1969). *Le conflit des interprétations. Essais d'herméneutique*. Le Seuil.
Ricoeur, P. (1974). *The conflict of interpretations: Essays in Hermeneutics* (W. Domingo et al., Trans., Don Ihde, Ed.). Northwestern University Press.
Taylor, C. (1975). *Hegel*. Cambridge University Press.
Thich, N. H. (1999). *The heart of the Buddha's teaching*. Three River Press.
Warnasuriya, K. S. (2007). Dependent origination and the Buddhist theory of relativity. *Hsi Lai Journal of Humanistic Buddhism (西來人間佛教學報), 8*, 153–163.
Zhuangzi, Z. (2009). *The essential writings* (B. Ziporyn, Trans.). Hackett Publishing Company.

Gerald Cipriani has taught philosophy for many years in Europe and East Asia, most recently at National University of Ireland Galway. Other positions include Visiting Professor at the Graduate Institute of Philosophy of National Central University Taiwan, Senior Lecturer at the Birmingham Institute of Art & Design, British Academy Fellow at Kyoto University, Visiting Scholar at the University of Helsinki, and Professor at Kyushu University, Japan. Invited lectures at a number of institutions including the University of Tokyo, Peking University, and University of London. He is also Honorary Professor to the UNESCO Chair in Comparative Studies of Spiritual Traditions, their Specific Cultures and Interreligious Dialogue. He studied Western aesthetics with full doctoral scholarship in Leeds (LMU) and East Asian philosophy and Kyoto School in London (SOAS) and Kyoto (KyotoU). He is the recipient of several research awards, including from National Science and Technology Council of Taiwan, Chinese Fund for the Humanities and Social Sciences, The Great Britain Sasakawa Foundation, The British Academy, and The Japan Society for the Promotion of Science. He is the chair of the International Research Group for Culture and Dialogue and Founding Editor of cross-cultural philosophy and humanities journal *Culture and Dialogue* (Brill) and *Journal of Aesthetics and Phenomenology* (Routledge). He has edited around 30 titles in philosophy and the arts and published numerous texts in the fields. His latest book publication is *Depois de Nishida: O Vazio Dialógico e a Cosmovisão Histórica* (After Nishida: Dialogical Emptiness and the Historical Worldview, trans. A. Bueno, 2023).

Recurring Dilemmas on Other Minds

Sanjit Chakraborty

Abstract This chapter critiques the notion of identity, a philosophical reaction in the Quinean sense that vindicates another significant hypothesis, which is called the "No entity without identity" theorem. Here, the identification of an entity relies upon its identity. The first segment of this chapter revisits the following question: How do we encounter the concept of a subjective "mind"? Could we "infer" or "experience" the concept of "mind" in our daily lives? Every entity has some characteristics that prompt comparison with similar entities. Although I think that other minds have a subjective periphery and that other minds have an objective periphery lurking behind this (privileged access), we can construe the relationship in a subjective way (individual mind). In the second section of the chapter, I address a different point by considering why the idea of "other minds" causes philosophical dilemmas. Can "other minds" exist as a substance in one's own mind? The category of human knowledge encompasses the reality of minds and other minds. The other minds remain an accompanying part of the individual mind since the expression, belief, and mental content of the individual mind merge in the linguistic expression in a split manner that we may ascribe as privileged access (subjective incorrigible knowledge) and corrigible knowledge. But I think that the relation between minds and other minds sounds especially idiosyncratic in its own identity. Despite becoming diverse, the individual mind can still be intimate by blending and attunement between the subjective and objective dichotomy.

Keywords Other minds · Dilemmas · Cartesian interactionism · Agent-relative knowledge · Behaviorism · Wittgenstein · Corrigible knowledge

Dilemmas are not a deplorable aspect of literature, but a theoretical conflict that aims to find a relevant solution in our philosophical sphere. Ryle depicts, "The internal administration of each seems to be impeccable but their diplomatic relations

S. Chakraborty (✉)
School of Social Sciences and Humanities, Vellore Institute of Technology-AP University, Amaravati, Andhra Pradesh, India

S. Chakraborty (ed.), *Human Minds and Cultures*,
https://doi.org/10.1007/978-3-031-56448-2_10

with one another seem to be internecine" (Gilbert, 1975, 1). The static part of the debate confined itself to the sphere of the contrast between common sense and scientific theory, which rather infringes on the catastrophic understanding of the nonlinguistic analysis of the mind. Dilemmas occur when we identify a particular entity, such as one's own mind and other minds, based on two or more contradictory criteria. Dilemmas, as I understand them, are the remnants of language that construct propositions that can be interpreted as a conflict between theories and platitudes. The chapter intends to explain the dilemmas surrounding the concept of other minds, a major debated issue in the current literature on the philosophy of mind.

Triggering Mind

The key question that defines the pendulum is the collapse of mind-body dualism. Cartesian interactionism preserves the amalgam of the mind and the body, although the mind is nonphysical, while the body does not have mental phenomena. Cartesian believes in the mode of the corporal body that could be extended (objectively) in limit but is inept at accounting for the mental states and understanding of the consciousness. So, it has become essential to invoke a second substance that seizes the nature of thought to explore the creative aspect of language and its use. The divisibility of the body and the indivisibility of the mind nudges a dualism in the Cartesian method. The interface could be possible between the mental and the physical states because of the *pineal gland* as Descartes portrayed unscientifically. This Cartesian myth hails the inner-outer problem in an unscientific way because of the theoretical cum categorical mistakes of putting the mind-body intersegment that Ryle later calls "ghosts harnessed to machines" (Ryle, 1976, 21). Debunking the mind-body dualism extends the psychological description of the mental states, while the body itself signifies the physical states. We can relook at the concept of mind from two different ways—functional and mental. Epiphenomenalism, a version of property dualism (or halfway to materialism) indicates human behaviors to the brain states and the causal effects on the objective world. Whatever may occur in the universe including the mental states has its physical causes, so we cannot rule out the physical states of the brain that causally entailed the mental phenomena or epiphenomena like desires, decisions, beliefs, etc.

A significant discrepancy between dualism and epiphenomenalism comes forward when dualists believe in the mental events that are located independent of physical events and their interaction may be feasible owing to interactionism where our mental events contain causal effects on the physical states or brains. Besides, epiphenomenalism believes in the process of mental events that are caused by physical events but does not believe in mental events, which affect the physical events; so, for them, believing the concept of other minds sounds inappropriate here. Epiphenomenalism rebuffs the physical effects of our mental events; thus, one has to grab either an inability to know our mental events or the knowledge of them could not be what may cause manifestly the physical event of saying something about the mental events. Cartesian moves toward a causal entwine between the mind and the material world by synchronizing the conviction of the causal "self-contentedness" pertaining to the objective world. The *by-product method* of epiphenomenalism involves mental phenomena being the side effects of objective phenomena, which have physical effects. The causal relationship between mental events and our nervous systems is rigid. The causal relationship between physical and mental phenomena seems ambiguous due to the necessity of overlooking the metaphysical reliance when reviewing the causal efficacy. Frank Jackson and David Michael wistfully write, "If dualists are committed to epiphenomenalism, then they are committed to saying that the very feature that makes something a desire for coffee on their view is causally irrelevant to the movement towards coffee. Similarly, the very feature that makes something an intention to get married or an itch is irrelevant to signing the marriage papers or scratching" (Jackson & Michael, 2007, 12).

Second, the idea of functional propensity as an anti-reductionist outlook certainly supervenes the mental states. Functional aptitude would not be a behavioral organism that is unintelligible beyond the boundaries of physicalism. Behaviorism does not staunchly appreciate any functional specification that articulates genuine causal efficacious internal states whose causal functions outweigh the *total functional specification* in general. Behaviorism posits a non-intentional calculative device that accurately connects external stimuli and behavioral exposition reflexively. Physical explanation, I think, cannot be the mere behavioral aspect because we cannot abandon intention that makes a bridge between explaining behavior and introducing the mental life. Physicalism derives from the argument of the possible states of the world while the mental states assess the argument of knowledge as it shuns the "no person theory" by defending the subjective agency of mental representation. The Cartesian model continues to treat the first person beyond the method of doubt. Anscombe looks at the first person or agency as self (an ontological understanding of the first person), which in our linguistic communication stands for "I." Now the question is whether the term "I" is a proper name, demonstrative, or referring expression. Anscombe (1977, 45–65) clarifies that it looks hard to consider "I" as a proper name since everyone uses it for themselves. The prime characteristic of a proper name merges individuality within a society where everyone uses the same name to identify the same reference. Besides, the demonstrative form of "I" may not be viable as an indexical demonstrative like "this or that," which could exemplify a proper name in a logical or linguistic function. The idea of using "I" as a singular demonstrative sounds intriguing. The contrast is that in the case of "I," the existence of the conscious directly secures its reference fixation, however, in most cases, demonstratives are failed to execute its reference like in the case of "this severe pain" or "that beautiful moment," etc. Even if it seems undeniably too hard to make out "I" as a refereeing expression since the inevitable part of the self-assertion does not represent the self. The myth of solipsism is only disproved by its internal contradiction due to its inevitable nature. When a person claims that "I am Sallu," it means that "She is not Sree." This statement proves the existence of other minds. Even one can extend their minds to the objects; here we can think about the automata and computers, etc. Karl Popper writes, "Einstein once said: 'My pencil is cleaver than I'. What he meant, of course, was that he could, by using his pencil, get results that he had not foreseen. Quite right: this is precisely the reason why we make pencils and use them. If we could not get further with a pencil than without it, we should not use it" (Popper, 1994, 109).

To draw the completeness of the world, Popper combats the physical world along with his theory of *world 3* and says that *world 1* discloses the physical states, while *world 2* affirms the disposition of behaviors. His argument against *world 3* is a way to explore the theories and objective problems. Popper argues that *world 1* is open toward *world 3* and *world 2* acts as a liaison. Popper drives an evolutionary approach to minds to get rid of epiphenomenalism (mind or consciousness seems like an epiphenomenon). Popper's new conjectural theory of mind defends a general evolutionary milieu and talks about interactionism between mental states and physical states in the sense of controlling systems and controlled systems. An agent's thought process incurs that we are conscious beings and a difference commences between the human consciousness and the lower-level consciousness of animals. Popper says, "Full consciousness is anchored in world 3—that is, it is closely linked with the world of human language and theories. It consists mainly of thought processes. But there are no thought processes without thought contents, and thought contents belong to world 3" (Popper, 1994, 114). Such an approach would be persuasive and here we find a way toward the cognitive demonstrative thought that illustrates a metaphysical dependency of the mind to the world like an agent's mind(s) will have a thought about an object (p) provided only the content of the thought, i.e., the particular object (p) exists and the agent's mind (s) is properly interrelated to the particular object (p).

Even mentioning a demonstrative object in the manner "this RW Roamer timepiece is an antique" identifies with the exact object (antique watch) that the agent's mind interacts with and perceives in a specific way, which is the way that the subject's mind can access the specific watch as an antique piece. There wouldn't be any thoughts if the subjects had no objects in their minds. Thus, the subject's thinking and the object have a causal relationship, and our mind mediates this linguistic game by giving it a conceptual milieu. Mentalists may believe that a person's mind is solely personal and is associated with their initial mental condition. A subject is thereby confined to their personal experience. The notions of self-restraint, self-emergence, or mental representation anticipate the person's motivational and intentional circuits. While the passive objective mode illustrates the content of the subject's thought, the active subject mode places the subject in the first person and makes them the authority of their acts of thought. This is because the term "I" sounds objective since it is associated with the subject's thought patterns and physical behaviors.

Davidson's model of triggering the mind deploys the causal power of mental contents like beliefs, desires, intentions, hopes, propositional attitudes, etc. What sorts of mental transactions do one hold between one's own mind and the mental contents? Davidson undermines a constituted austere sense of the normative outlook on the mind, where mental states are the causal dispositional attitudes to behave in a certain way which would be only answerable to the principle of rationality. Mental contents like beliefs, desires, hopes, etc. are not actually in any norm that Wittgenstein preserves in his writings. For Davidson, these are considered to be casual dispositions. Davidson prioritizes the only norm of the mind which is nonetheless the principle of rationality. Bilgrami writes, "Davidson thinks that our states of mind such as beliefs, desires, and intentions are not themselves commitments but rather (causal) dispositions which are answerable to the only real commitments we have—to the principles of rationality such as consistency, transitivity, etc. By contrast, Wittgenstein thinks that our states of mind are commitments or internal norms. Our mere possessions of such states of mind are *commitments* to doing and thinking various things. So, for Wittgenstein, the mind is cluttered with far more norms and commitments than it is for Davidson. Davidson's view of norms is an austere one (the only commitments we have are our commitments to the principles of rationality such as consistency, transitivity, etc., all other states of mind are dispositions). Wittgenstein's is a more bloated view (each and every belief, desire, and intention is itself a commitment)" (Bilgrami, 2017, 76).

In Wittgenstein's *Tractatus,* we find a position in which the solipsistic account goes with realism, seeing as everything in the world is equally real, from the mountain to our mental states. But all these are inseparable parts of the subject's objective world since the subject (I) objectively describes the world in linguistic communication. Being a thinking agent, everything in the world is an expression of the subject's language and words. Eventually, this turns into an egocentric predicament, which Wittgenstein tries to sidestep in his later writings. The subjective identification of one's own pain or pleasure looks self-regulating and self-contained, but Wittgenstein puts it differently by arguing that this self-contained analysis is nothing but a presupposition of the integrity of the personal line which relies on the abstraction of language from the sphere of external manifestation and sense-experience that edifies the ascription of language to others in our linguistic community. Thomas Nagel urges, "Wittgenstein does not offer an analysis of meaning, but he does try to identify certain attempts to use language that cause it definitely to lose its moorings, so that the distinction between correct and incorrect application of the term in question no longer exists, and all we have left is the *word* (usually in italics). He believed that the mythical private language of experience, completely detached from anything in our behavior and circumstances that could be observed by others, which is used to express radical skepticism about the experience of others, is an example of this. We fail to see it because we unconsciously rely on our understanding of the experiential concepts of the public language, whose conditions of meaning are concealed in their subjective application to ourselves" (Nagel, 1995, 51).

It is worth remarking that the private language argument advocates for the solitary, meaningful use of language and words without any interaction with others. Here the argument of an analogical inference may be unacceptable in this fashion where other minds preserve a possibility that helps to understand the correct or incorrect use of language through the checked-up model of user's practices. Supporters of private language arguments could argue against the asymmetry, which can cause difficulties when checking users' practices for correctness or incorrectness. Epistemologically, the private language argument and the problem of other minds are two discrete problems that cope with two different knowledge systems—corrigible and incorrigible knowledge, where the scopes and limits of the argumentation of these two camps are clarified. If we mold the problem of private language midst of the private sensation or the internal states of mind, undoubtedly the knowledge of the individual may stipulate the first-person incorrigibility about this particular sensation (let us think that it is about one's itching) that no one else apart from the individual could experience directly the same sensation of itching. The quale of mental state or sensation is nothing but the introspective phenomenal characteristic that represents only the conscious state of the person, which neither functionalism nor behaviorism are compatible with and beset in the tune of intrinsic states of the mind as well. In short, a formidable array of private language arguments relies on the query of whether the meaning could exist devoid of the profligacy in the checking method of its correct or incorrect uses.

An Omnibus of Other Minds

The epistemic asymmetry that talks about the privileged access and the first-person authority of the speaker can be challenged by others who consider that one can know the subject's mind better than the subject. All these efforts are unwilling to inculcate the essence of the mind as inner or private. The analytic trend of the twentieth century shifts toward externalism, which consists of meanings that are publicly shareable and also rely on the sociolinguistic paradigm. They reject the notion of intrinsic meaning being found in minds as a museum myth. The critics argue that the subject is the first-person authority in their private world, especially what is going on inside the mind, they can be directly aware of their mental states. To observe an agent's behaviors, a third person can only infer or imagine these things. The behaviorist's way of seeing the mind as something outer tries to reject the ontological innerness by bringing the idea of epistemology where one can resolve the external determining items that underscore the framework of an agent's beliefs on holding some prescriptive roles. It would be intriguing, as Colin McGinn claimed (Colin, 2017, 57–58), to think of the mind as both internal and external, with outer aspects like intention, action, and effects. All these criteria of the mind have some tendencies to engage with the external world. I need to explain one more point, however, a precise opinion cannot be formed when the naturalistic outer world denies the mind as an intrinsic state. The representation of mental states can be achieved by discussing the mind and other minds. The mind is something which can only be studied by introspection. The agent's thoughts, imaginations, and emotions are not limited to communication, so we can place them in the realm of inner entities that the subject has privileged access to. To me, the module of innerness of mind is viewed as an intrinsic part of the brain. It can only be expressed in the public sphere when a subject structures their thoughts or sensations in a sentence to confer it a linguistic form that could be shareable with others in our sociolinguistic communication.

Now the problem is "Do we consider that a sort of 'constitutional uncertainty' plays a germane role in our phenomenal reports (inner) as well as in our sociolinguistic behavior (outer)?" While incorrigible reports may contain information about causal properties that have a causal effect on behaviors, language allows us to communicate publicly about phenomenal reports. Minds are intermingled with others as the content of one's thoughts can be articulated in propositions. The mental states or thoughts are subjective (in the sense of privileged access), but the content of our thoughts turns toward objectivity as it loses its intrinsic privacy when it is going to be shared with others in any mode of linguistic communication. Language allows us to put across expressions and concepts with others since the uniformity of alliance of words that infuse concepts could have been learned and communicated through other minds. So, our minds (in plural) are embedded in the world and causally interacting with the world (objective world, no other possible worlds) in a way that the internal states represent the way the world is (not the way the world would be). Furthermore, we (the agents) use words to convey to others how our mental states or thoughts represent things. The concept caters to an ostensible role here. The concepts aid us in comprehending the impact of the world on us. Even the causal history of content involves a shared language, and our place in a society is as language users who use or practice content that may be intersubjective. There are perceptual concepts, sensations, and emotions that I believe are beyond language use. Thoughts and all these may be internal in the sense that the agents have the first-person authority about these, but their contents can be understood or judged only if they are described in language (Chakraborty, 2020).

Let us take a skeptical look at the debate of other minds. The following two sentences can be pinned down in a similar way:

(a) Other people have their minds.
(b) Other people have their itching.

In a sentence, if an agent asserts that "other individuals possess their thoughts," it implies that the agent holds the belief in the speculation of other individuals. The agent holds the belief that "other people have their itching" and believes that the feeling of itching is accessible to all. The inquiry now is, how could the agent address the issue of itching? Now, the query is, how could the particular agent know about this knowledge? Let us understand a behavioral appeal. If I argue that other people have the sensation of headache as other people have minds, the progression of my knowledge about the particular claim depends on the retrospection of *knowing that* method. I think this *knowing that* method of explaining other minds and others' headaches adjoins an inductive justification of securing behaviorism from the scale of psychological analysis. One can avail the knowledge of other minds just by comparing the behavioral dispositions of others' mental states like anxiety, anger, love, hate, etc. since my mind preserves the same mental states, and most of the time, my behavior expresses the actions that my mental states structured. So, in a specific way, the mental states of other people and their behavioral patterns explore the presence of other minds. One thing needs to be clear here that we are talking about other minds hereby excluding animal minds, as Hilary Putnam beautifully argues against Paul Ziff that "What he (Ziff) says is that if I assume that others have minds, and that 'others' includes animals (rats, etc.), and that the mind and brain stand in a significant relation, and to these hypotheses, I conjoin 'still others', which he does not state, then I obtain a conceptual scheme which 'draws support from a multitude of observations and experiments'" (Putnam, 1986, 347).

How we could learn to employ the phrase "other minds" in language is an important finding. Is it a psychological prediction regarding other people's actions? The way people behave validates that an agent may infer information about other's mental states from their outward displays of affection, kindness, anger, etc. The peculiar aspect of this language focuses on what "angry" and "love" stand for. Here, the agent must understand what is meant by the terms "angry" and "love," or else it will be difficult for them to understand the specific mental states of the person they would like to refer to as an "angry young man." And the progress undoubtedly seems doable just because of the knowledge of the particular meaning of the term "angry." The behavioral indicator sophistically endures with linguistic indulgence. In order for there to be any successful communication between the hearer and the speaker, the idea of sameness of meaning may have an interesting role in symbolizing the behavioral indicator. In this context, Putnam writes, "If I say of someone who grimaces that he is happy, this is because I believe that the state that I call *happiness* is one that *could* lead to that behavior under *some* circumstances and that those circumstances obtain in the case in question. If it were deeply embedded in my psychological theory that under no circumstances can a person who grimaces be happy, then I might be disinclined to say that the meaning of 'happy' was the same even in the culture described a few moments ago in our culture...It only follows that a given behavioral sign can be a sign of happiness in some circumstances only if the people who take it to be a behavioral sign in those circumstances have appropriate other beliefs, habits, customs., in connection with happiness" (Putnam, 1986, 352).

The private language argument, which rejects the behaviorist explanation in favor of the individual mind, can be contested here. The cohesiveness of personal experience that linguistic behaviorism uses appears erroneous and undeserving. We do not have private access (incorrigible experience) to our mental and physical states, even though we can discuss them. When we look for evidence of other minds, we run into issues. Do any other minds actually exist? Then, the crucial question is, "Is it possible for an individual mind to acquire a specific language rule that might not be shared with other minds?" I believe that the speaker's interpretation of what a referendum is cannot constitute the definition of a referendum in the context of a common language. A definition needs to be validated by the other minds or else the definition could go wrong, and the understanding level of the particular definition of the referendum would be restricted to the domain of the solipsistic stage. To clarify Wittgenstein's view, Rhees argues, "Wittgenstein did not say that the ascription of meaning to a sign is something that needs justification. That would genuinely be as meaningless as it would be if you said that language needs justification. What Wittgenstein did hold was that if a sign has a meaning it can be used wrongly. On the other hand, if anyone had tried to invent language and teach it to others, you might say the language and the use of expressions did stand in need of justification" (Rhees, 1971, 68). There is no logical travesty to claim that the meaning of something and its conceptual facet may be conceived through the association with other minds. The idea of misunderstanding bluntly belongs to the part of interacting with other minds. Saying anything about language acquisition implies that it is not a mental game played alone by the individual in any way. Even though someone has a private code, they do not have a private understanding because understanding must be expressed in a language that can be shared with others.

Dilemmas and Way Out

In the field of epistemic inquiry, the issue of other minds poses a skeptical challenge by examining how one can ascertain the existence of another person's mental states. Here, rather than focusing on the conceptual structure of the problem, I would consider the epistemic concern raised by these conundrums. Philosophical curiosity about the existence of other minds is known to exercise people's mental states and systematic thought procedures in first- and third-person authoritative contexts. This is a prima facie issue about the obstruction of others' mental states that also rests on the barrier of separation, i.e., separating one's own mind from that of others (Strawson, 1977, 99–100). Strawson simplifies the idea of other minds by putting it into the sphere of other ascription in addition to the self-ascription. This principle caters a foul against the analogical inference as one cannot stir from one's sensations and coalition with others. According to Strawson, an individual cannot talk about one's own experiences unless one is acquainted with how to assign or ascribe experiences to others. Strawson writes, "There is no sense in the idea of ascribing states of consciousness to oneself, or at all, unless the ascriber already knows how to ascribe at least some states of consciousness to others. So, he cannot argue in general 'from his own case' to conclusions about how to do this; for unless he already knows how to do this, he has no conception of *his own case*, or any *case*, i.e., any subject of experiences" (Strawson, 1977, 106). The critical assertions genuinely challenge the thesis of analogical inference by relying on an argument in favor of other minds. The abridgment thesis now outlines a variety of distinct persons that clarify a predicative ascription to several distinct agents; in other words, the predicate must be attributed to multiple agents or individuals. Additionally, Strawson's second claim discusses epistemic logical adequacy because logical adequacy aids in avoiding skeptic subtleties, and the concept of adequacy directly nurtures an epistemological sense of rational belief in one's statement. According to the second claim, when someone meaningfully attributes anything to themselves or others in particular, they must possess a magical ability that carries over completely different criteria.

Mind secludes the principle of individuation while other minds pivot the creation of the identity of minds. The discussion about the notions of mind and other minds construes the specific object that we call *self-notion* or, in John Perry's tune, the

"Agent-relative knowledge," where the subject's mind is intimately and inseparably engaged in the notion of the self-thought like "I am a philosopher." However, since the subject has both first-person authority of their mind and conceptual access to the minds of others, one could argue that a general idea of their mental state may be possible.[1] We need to have a meaningful understanding of our mental states and the ideas we support before we can extend the mental replacement to others. Those who contend that knowledge of other minds comes from an inferential process devoid of primordial experience are presenting an incorrect picture of the problem. An agent may possess innate-based or first-person access to their thoughts differently. The tacit knowledge could be expressed in the language where the subject expresses their mental states and ascribes the same mental states to others. To get a justification for the subject's mental beliefs, we have to roll on the epistemic access of others' mental beliefs. Here, the idea of the justification of the first-order belief does not cope with the self-knowledge of the agent's mind from a conceptual stance but invades the resolution of the knowledge claim to undergo a successful communiqué with others. An agent has a striking ability to classify oneself as an agent of a particular action and think up a distinction between self-generated actions with others generated actions.

The unique part of the mind elucidates two different ascriptions—an inner awareness of the mental states of the individual and an outer manifestation of these mental states in behavior or speech acts. A subject-predicate form is required to structure the content of an agent's belief, with the ascription emphasizing the predicate that synchronizes the belief's content. Using the same example, "I am a philosopher," the person making the attribution (the subject of the attribution) would appropriately claim the attribution for themselves. The ascription seems to be a reflexive ascription of the speaker's (the ascriber's) mental states. However, it's possible that an assertive account from a first-person authority could be misreported due to self-deception or inaccurate reporting, such as saying things like "I am an incarnation of God," etc. Kirk Ludwig points out, "First-person knowledge and first-person authority are not equivalent. First-person knowledge does not necessarily require first-person authority, because a difference between the way one knows something and the way others know it does not entail that there is a difference between the authorities with whom one knows. And first-person authority does not by itself imply that there is a difference in the way one knows one's mental state. Nonetheless, one should expect there to be some intimate connection between first-person knowledge and authority" (Ludwig, 1994, 371).

If we look at the behaviorist or inferential account, an agent's behavior sometimes barely entails the mental states of the person since we can pretend our mental

---

[1] Let me clarify here that Dennett holds a wonderful position and writes, "As creatures of our own attempts to make sense of ourselves, the putative mental activities of folk theory are hardly a neutral field of events and processes to which we can resort for explanations when the normative demands of the intentional system theory run afoul of a bit of irrationality" (Dennett, 1995, 190). I agree with him about this startling and relevant discussion, but here my true concern is to understand how we could acquire the development to get involved in it.

states by exposing a different behavior. The dilemma equips a hassle between the articulation of the mental states and exposing a different behavior. But Quine critically argues that "Correspondingly, instead of saying that mental states are identical with psychological ones, we could repudiate them; we could claim that they can be dispensed with, in all our theorizing, in favor of psychological states, these being specified usually not in actual psychological terms but in the idiom of behavioral dispositions. This repudiation version has a certain advantage, though a different one from what we noted in the case of number. Its advantage here is that it discourages a possible abuse of the identity theory. For product through the identity theory is of hard-headed materialism, we must beware of its sedative use to relieve intellectual discomfort" (Quine, 1977, 94–95).

The basic purpose of this research is to investigate the puzzle of the mind and other minds' relationship. Although an understanding of how other minds convey information about access presentation pivots a dispositional attitude that functions in intransitive and informative ways in the language of thought, the individual mind's mental state stands for an occurrence since it has an incorrigible cognitive access. Just to nail the problem down a bit more precisely, there is an interesting picture induced from the dictum that knowledge of own mind is logically and conceptually prior to knowledge of other minds. The paradigm of an individual mind is sensational while other minds conceptualized its existence in language by bringing the notion of *that clause*. The dilemma persists when we query the existence of an individual mind by discarding other minds. Can the mind exist without being clustered with other minds?

The twofold criterion of the mind exclusively draws a picture where the essential part is sensation, and the second part consists of propositional contents using *that* clause. The sensational part of the mind looks primitive where an agent feels something from the distinctive phenomenological access. However, the conception of propositional contents has a prerequisite that overlies the sensational part of the mind since the rational appeal of the agent construed the dispositional properties of the mental states in language. So here the rational form and language play a pertinent role. Colin McGinn raises a wonderful analogy between the visual and textual sequences on one side, and on the other, he urges first-person and third-person perspectives. He thinks that in each case one has to be abreast of a mode of presentation. McGinn writes, "The manner in which normative considerations operate from the third-person point of view is mirrored in a certain way from the first-person perspective, and this brings out a connection between rationality and self-consciousness" (McGinn, 1996, 21). The mind is a mongrel concept. I think the concepts of consciousness and intentionality are the ontological criteria that could be regarded as inseparable characteristics of the mind. The mind being an intrinsic and incorrigible sensation of our mental states holds the first-person ascription in the case of the agent, but the same mind looks extrinsic to the propositional contents or language use because of the deflective part of third-person ascription in our linguistic communication. The subjective states of intentionality direct toward the inner mind and external objects equally, and the whole process endures consciousness in a spinning way (causally intertwined). Simultaneously, the mental phenomena encounter subjectivity (cognition), which may constitute an intriguing aspect of the epistemological discourse to elucidate their source. As beliefs and mental contents of the ascriber's mind may blend in the linguistic expression in the sense of a split act, the other minds essentially remain an accompanying component of the individual mind. This could be assigned to privileged access (subjective incorrigible access) and a corrigible knowledge system. This illustrates how fascinating the relationship is between the mind and other minds.

How can we know other minds epistemically? An overview of knowledge describes how one obtains knowledge from several sources, including observation, inference, testimony, and introspection. Since there is no certain way to learn about other minds, it appears to be a little challenging to understand other minds and their emotions. There are numerous similarities and dissimilarities available between the behaviors and life forms of the fakir, Baul (the mystic singer of Bengal), eccentrics, and mad people, making their feelings difficult to comprehend. To understand an agent's thought patterns, we must maintain a long-term relationship of intimacy and acquaintanceship. Well, let us consider that we have this sort of prospect in our daily life. Here, the point is that how could it be possible to know the feelings of other minds or what another person is feeling or thinking about?

Essentially, the response is that individuals can infer that other people have minds similar to ours, which allows them to think and feel like we do simply by seeing their actions, facial expressions, and physical changes. Second, the next method that might help us uncover the existence of other minds is the process of inference. An agent infers an epistemic possibility of knowing other minds and others' contents of thought because the agent's mind might help them to think or feel about these. Others also have minds that enable them to feel and think similarly. The model of testimony is another source of knowledge that could assist us with the other minds in retrospect of the other's utterances and expressions. Because introspection bestows privilege upon the subject from a satisfyingly incorrigible base that sketches the other minds according to introspective access to one's own concept, it sounds like a method of knowing other minds. These are the sources (perception, inference, testimony, and introspection) that deal with other minds and their mental accessibility. According to the skeptics, as the problems of the sources do not have a pleasing way out, there is no doable question of the existence of other minds in our literature.

There is a promising way to relook at other minds' problems from the elevation of a nonempirical stance. The nonempirical stand initiates a priori knowledge of the existence of other minds. But the perceptual model is the best source of our knowledge and the model of inference, testimony or a priority could not infringe a ground argument to get rid of the problem of other minds. The perceptual model of other minds develops a source of epistemic justification for reasonable thought procedures of other minds. Cassam writes, "The position is not that we first have to come up with a justification for thinking that there are other minds and then figure out what others are thinking and feeling. Rather, we can come to know *that* there are other minds by seeing *what* others are thinking and feeling" (Cassam, 2009, 159). The perceptual model depends on our ability to understand the thoughts of others by seeing their facial expressions, behaviors, and symbols when they are depressed, angry, or in any other emotional mood. One challenge facing proponents of the perceptual model of other minds is that we are only able to watch the outward manifestation of anger in an individual; we are never able to observe and understand the notion of anger itself. Since there is no set standard for identifying our emotional states, interpreting signs or expressions of anger appears to be only an interpretive process that can be modified. Outlining a behaviorist stance, one could claim that the observable aspects of their changes in physical characteristics, such as their eye expression, facial reddening, body language, and speech patterns, are examples of physical changes. These alterations are thought to be the literal manifestation of a certain level of escalation of anger. One may argue, however, that our perception of what we take to be angry expressions is no more significant than our perception of a real object—a burning iron rod, for example. Could both physical objects and human emotional expressions benefit from the externally based changes? A hoax that no one could avoid here is the notion of mental analogy, which has a self-deception mode, and the reader can easily verify that no unconscious object ever attains it. Despite the biological and physical similarities, the epistemic access allows us to believe that we may have similar minds if we rule out the possibility that the person, we met is not an automaton or zombie but rather a person with a mental life form (the identity of the mental states and its behavioral exposures). Furthermore, in this instance, we comprehend the concept of attributing mental states to other minds. In addition to a self-ascription mode, the identity condition of the mental states reveals additional ascription aptitudes (other ascription) in support of comprehending the other minds. The identity condition's conception comes to be quasi-empirical (an anti-minimalist account), as its construction originates at the conceptual apparatus level, which preserves nonempirical a priori knowledge and its subsequent derivation, such as reason-based intuitive knowledge on the one hand and a language-centric representation-based module on the other. In an expanded framework of philosophical analysis, encoding linguistic representation and conceptual apparatus in an intuitive sense may outline a trajectory for the quasi-empirical hypothesis that integrates the mind with other minds.

References

Anscombe, E. W. S. (1977). First person. In S. Guttenplan (Ed.), *Mind and language*. Clarendon Press.

Bilgrami, A. (2017). Norm and failure in mind and meaning. *Argumenta, 3*(1), 69–85.

Cassam, Q. (2009). *The possibility of knowledge*. Clarendon Press.

Chakraborty, S. (2020). *The labyrinth of mind and world: Beyond internalism-externalism*. Routledge.

Dennett, D. C. (1995). Making sense of ourselves. In W. G. Lycan (Ed.), *Mind and cognation a reader*. Blackwell.

Gilbert, R. (1975). *Dilemmas*. Cambridge University Press.

Gilbert, R. (1976). *The concept of mind*. Penguin Books.

Jackson, F., & Mitchell, D. B. (2007). *Philosophy of mind and cognition*. Blackwell Publishing.
Ludwig, K. A. (1994). First-person knowledge and authority. In G. Prayer, F. Siebelt, & A. Ulfig (Eds.), *Language, mind, and epistemology: On Donald Davidson's philosophy*. Kluwer Academic Publishers.
McGinn, C. (1996). *The character of mind*. Oxford University Press.
McGinn, C. (2017). *The philosophical provocations*. MIT Press.
Nagel, T. (1995). Wittgenstein: The egocentric predicament. In *Other minds: Critical essays 1969–1994*. Oxford University Press.
Popper, K. (1994). *Knowledge and the body-mind problem*. Routledge.
Putnam, H. (1986). Other minds. In *Mind, language and reality* (Vol. 2). Cambridge University Press.
Quine, W. V. (1977). Mind and verbal dispositions. In S. Guttenplan (Ed.), *Mind and language*. Clarendon Press.
Rhees, R. (1971). Could language be invented by a Robinson Crusoe. In O. R. Jones (Ed.), *The private language argument*. Macmillan St Martin's Press.
Strawson, P. (1977). *Individuals*. Methuen.

Sanjit Chakraborty is an Assistant Professor in the School of Social Sciences and Humanities at Vellore Institute of Technology-AP University. Before that, he taught at the Indian Institute of Science Education and Research Kolkata, the Indian Institute of Management Indore, and the Central University of Hyderabad. His philosophical venture was nourished under the guidance of Professor Hilary Putnam (Emeritus Professor, Harvard University) from 2008 to 2016. His work spans the topics of philosophy of mind, philosophy of language, philosophy of science, ethics and AI and morality. Chakraborty's books include *Engaging Putnam* (De Gruyter, 2022), *Living Without God: A Multicultural Spectrum of Atheism* (Springer, 2022), *The Labyrinth of Mind and World: Beyond Internalism-Externalism* (Routledge, 2020), *Understanding Meaning and World: A Relook on Semantic Externalism* (CSP, 2016), and *Pursuits of Language* (forthcoming). Chakraborty has extensively published many papers in much-respected peer-reviewed international journals, and his works have been reviewed and cited in reputed international journals by noted scholars. Chakraborty has been invited for talks at different renowned institutions and universities overseas. Chakraborty holds the editor and referee positions in the journals of *Human Affairs* (De Gruyter), *Journal of Scholarly Publishing* (University of Toronto Press), *Journal of Experimental & Theoretical Artificial Intelligence* (Taylor and Francis), *Culture and Dialogue* (Brill), *Research Ethics* (Sage), and *Sophia* (Springer) and book reviewer of Springer, Routledge, World Scientific Publishing, and Palgrave Macmillan.

References

Adorno, T. W. (1984). *Aesthetic theory* (R. Hullot-Kentor, Trans.). Continuum.

Ambrose, S. E. (2002). The Last Barrier. In R. Cowley (ed.), *No end save victory: New second world war writing*. Putnam Adult.

Andrew, J. P. (2021). *Freedom, mattering, and the intrinsic badness of suffering* (dissertation). University of Texas at Austin.

Anscombe, E. W. S. (1977). First person. In S. Guttenplan (Ed.), *Mind and language*. Clarendon Press.

Ardrey, R. (1961). *African genesis: A personal investigation into the animal origins and nature of man*. Atheneum.

Arpaly, N. (2003). *Unprincipled virtue*. Oxford University Press.

Assembly, U. G. (2007). United Nations declaration on the rights of indigenous peoples. *UN Wash, 12*, 1–18.

Bang, J. P. (1917). *Hurrah and hallelujah: The teaching of Germany's poets, prophets, professors and preachers*. George H. Doran.

Barber, A. (2013). Science's immunity to moral refutation. *Australasian Journal of Philosophy, 91*(4), 633–653.

Basler, R. (Ed.). (1953). *The collected works of Abraham Lincoln* (Vol. 8). Rutgers.

Baxter, H. (2011). *Habermas: The discourse theory of law and democracy (jurists: Profiles in legal theory.)*. Stanford Law Books.

Benedict, R. (1934). A defense of cultural relativism. *Journal of General Psychology, 10*, 59–82.

Berkowitz, D., Pistor, K., & Richard, J. F. (2003). The transplant effect. *American Journal of Comparative Law, 51*(1).

Bernstein, R. J. (1997). *The restructuring of social and political theory*. Methuen & Co Ltd.

Bertrand, O. (2016). Reducing suffering or adding years of life? *14th conference of the international society for Utilitarian studies*.

Bilgrami, A. (2017). Noam and failure in mind and meaning. *Argumentation, 3*(1), 69–85.

Bloom, P. (2013). *Just Babies*. Crown.

Bolton, R. (2005). *Habermas's theory of communicative action and the theory of social capital*.

Bourguignon, F. (2015). *The globalization of inequality* (T. Scott-Ralton, Trans.). Princeton University Press.

Brewer, M. B. (1999). The psychology of prejudice: Ingroup love or outgroup hate? *Journal of Social Issues, 55*, 429–444.

Brink, D. (1989). *Moral realism and the foundations of ethics* (pp. 197–209). Cambridge University Press.

S. Chakraborty (ed.), *Human Minds and Cultures*,
https://doi.org/10.1007/978-3-031-56448-2

Broi, A. (2022). Une défense de l'hédonisme axiologique. *Dialogue: Canadian Philosophical Review/Revue canadienne de philosophie, 61*(2), 325–346.

Bürger, P. (2004). *Theory of the Avant-Garde* (M. Shaw, Trans.). University of Minnesota Press.

Cassam, Q. (2009). *The possibility of knowledge*. Clarendon Press.

Ceci, S. J., & Williams, W. M. (2009). *The mathematics of sex: How biology and society conspire to limit talented women and girls*. Oxford University Press.

Chagnon, N. (1988). Life histories, blood revenge, and warfare in a tribal population. *Science, 239*, 985–992.

Chakraborty, S. (2020). *The labyrinth of mind and world: Beyond internalism- externalism*. Routledge.

Chakraborty, S. (2021a). Pursuits of belief: Reflecting on the cessation of belief. *Sophia, 60*(3), 639–654.

Chakraborty, S. (2021b). Spinning solitude: Coronavirus and the philosopher. *Sophia, 60*(3), 769–774.

Charles, G.-U., & Fuentes-Rohwer, L. (2015). Habermas, the public sphere, and the creation of a racial counter public. *Michigan Journal of Race and Law, 21*, 1.

Chomsky, N. (1955). *Semantic considerations in Grammar*, Monograph No. 8, Georgetown Monograph Series.

Comesaña, J. (2006). A well-founded solution to the generality problem. *Philosophical Studies, 129*, 27–47.

Conee, E., & Feldman, R. (1998). The generality problem for reliabilism. *Philosophical Studies, 89*, 1–29.

Cook, T. (1999). *No place to run: The Canadian corps and gas warfare in the first world war*. UBC.

COPE Council. (2022). *COPE discussion document: Diversity and inclusivity*. https://doi.org/10.24318/RL.qSoVsZ.

Crisp, R. (2006). *Reasons and the good* (pp. 107–109). Oxford University Press.

Crisp, R. (2022). Deep personal relationships and well-being: A response to hooker. *Ratio, 35*(4), 301–309.

Darwin, C. (1859a). *The origin of species*. John Murray.

Darwin, C. (1859b). *On the origin of species by means of natural selection, or the preservation of favoured races in the struggle for life*. John Murray.

Darwin, C. (1871). *The descent of man*. John Murray.

Darwin, C. (1998 [1872]). *The expression of emotion in man and animals*. HarperCollins.

David, R. (1950). Traité élémentaire de droit civil comparé 225.

de Lazari-Radek, K., & Singer, P. (2014). *The point of view of the universe: Sidgwick and contemporary ethics*. Oxford University Press.

de Schaepdrijver, S. 2014. The German atrocities of 1914. *British Library: World War One*, https://www.bl.uk/world-war-one/articles/civilian-atrocities-german-1914.

de Waal, F. (1996). *Good natured: The origins of right and wrong in humans and other animals*. Harvard University Press.

Decker, J., & Groll, D. (2013). On the (in)significance of moral disagreement for moral knowledge. In *Oxford studies in Metaethics* (Vol. 8). Oxford University Press.

Deleuze, G. (1969). *Logique du sens*. Les Éditions de Minuit.

Delon, N. (2022). Strangers to ourselves: A Nietzschean challenge to the badness of suffering. *Inquiry*, 1–30.

Dennett, D. C. (1995). Making sense of ourselves. In W. G. Lycan (Ed.), *Mind and cognation a reader*. Blackwell.

Dickens, C. ([1857] 1948). *Little Dorrit*. Oxford University Press.

Doris, J., & Plakias, A. (2008). How to argue about disagreement: Evaluative diversity and moral realism. In W. Sinnott-Armstrong (Ed.), *Moral psychology* (Vol. 2, pp. 303–331). MIT Press.

Driver, J. (2001). *Uneasy virtue*. Cambridge University Press.

Dukes, D., et al. (2021). The rise of Affectivism. *Nature Human Behaviour, 5*, 816–820.

Dyble, M., Salali, G. D., Chaudhary, N., Page, A., Smith, D., Thompson, J., Vinicius, L., Mace, R., & Migliano, A. B. (2015). Sex equality can explain the unique social structure of hunter-gatherer bands. *Science, 348*(6236), 796–798.

Elga, A. (2007). Reflection and disagreement. *Noûs, 41*(3), 478–502.

Enoch, D. (2009). How is moral disagreement a problem for realism? *The Journal of Ethics, 13*, 15–50.

Feldman, F. (2004). *Pleasure and the good life*. Oxford University Press.

Ferguson, N. (2004). Prisoner taking and prisoner killing in the age of total war: Towards a political economy of military defeat. *War in History, 11*(2), 148–192.

Ferguson, R. B. (2013a). Pinker's list: Exaggerating prehistory war mortality. In D. P. Fry (Ed.), *War, peace, human nature: The convergence of evolutionary and cultural views* (pp. 112–131). Oxford University Press.

Ferguson, R. B. (2013b). The prehistory of war and peace in Europe and the near east. In D. P. Fry (Ed.), *War, peace, and human nature: The convergence of evolutionary and cultural views* (pp. 191–240). Oxford University Press.

Ferguson, R. B. (2015). History, explanation, and war among the Yanomami: A response to Chagnon's *Noble savages*. *Anthropological Theory, 15*, 377–406.

Festinger, L. (1957). *A theory of cognitive dissonance*. Stanford University Press.

Finnis, J. (1980). *Natural law, natural rights*. Clarendon Press.

Ford, J. C. (1944). The morality of obliteration bombing. *Theological Studies, 5*, 261–309.

Forsyth, A. (2019). Law as science. In *Common law and natural law in America*. Cambridge University Press.

Foucault, M. (1977). *Discipline and punish: The birth of the prison* (A. M. Sheridan-Smith, Trans.). Vintage.

Frank, R. (1988). *Passions within reason: The strategic role of the emotions*. W. W. Norton.

Fraser, N. (1985). What's critical about critical theory? The case of Habermas and gender. *New German Critique, 1985*, 35.

Freud, S. (1916). *Leonardo da Vinci a psychosexual study of an infantile reminiscence* (A. A. Brill, Trans.). Moffat, Yard & Company.

Freud, S. (1920. A general introduction to psychoanalysis (G. Stanley Hall, Trans.). Horace Liveright Publisher.

Freud, S. (1999). *The interpretation of dreams* (J. Crick, Trans.). Oxford University Press.

Fry, D. P. (2013). War, peace, and human nature: The challenge of achieving scientific objectivity. In D. P. Fry (Ed.), *War, peace, and human nature: The convergence of evolutionary and cultural views* (pp. 1–21). Oxford University Press.

Fry, D. P. (2014). Group identity as an obstacle and catalyst of peace. In J. F. Leckman, C. Panter-Brick, & R. Salah (Eds.), *Pathways to peace: The transformative power of children and families* (pp. 79–92). MIT Press.

Garfield, J. L. (2001). Nagarjuna's theory of causality: Implications sacred and profane. *Philosophy East and West, 51*(4), 507–524.

Geertz, C. (1973). *The interpretation of cultures*. Basic Books.

Gensler, H. J. (1996). *Formal ethics*. Routledge.

Gensler, H. J. (2013). *Ethics and the golden rule*. Routledge.

Gensler, H. J. (2016). *Ethics and religion*. Cambridge.

Gensler, H. J. (2017). *Introduction to logic* (3rd ed.). Routledge.

Gensler, H. J. (2018). *Ethics: A contemporary introduction* (3rd ed.). Routledge.

Gilbert, R. (1975). *Dilemmas*. Cambridge at the University Press.

Gilbert, R. (1976). *The concept of mind*. Penguin Books.

Giroux, H. (2014). *Neoliberalism's war on higher education*. Haymarket Books.

Glaeser, E. L., & Shleifer, A. (2002). Legal origins. *Quarterly Journal of Economics, 117*(4), 1193–1229.

Glynn, P. (1997). *God: The evidence*. Prima.

Goldman, A. (2000). What is justified belief? In E. Sosa & J. Kim (Eds.), *Epistemology: An anthology*. Blackwell.
Goodall, J. (1986). *The chimpanzees of Gombe: Patterns of behavior*. Cambridge, Mass.
Grossman, D. (2009). *On killing: The psychological cost of learning to kill in war and society*. Back Bay Books.
Grotius, H. ([1625] 1901). *The rights of war and peace, including the law of nature and of Nations*. Editor A. M. Campbell. M. Walter Dunne.
Haas, J., & Piscitelli, M. (2014). The prehistory of warfare: Misled by ethnography. In D. P. Fry (Ed.), *War, peace, and human nature: The convergence of evolutionary and cultural views* (pp. 168–190). Oxford University Press.
Habermas, J. (1981). Modernity versus postmodernity. *New German Critique, 22*, 3–14.
Habermas, J. (1984). *The theory of communicative action*, Vol. 1 (T. McCarthy, Trans.). Beacon Press.
Habermas, J. (1987). *The theory of communicative action*, Vol. 2 (T. McCarthy, Trans.). Beacon Press.
Habermas, J. (1996). Three normative models of democracy. In S. Benhabib (Ed.), *Democracy and difference: Contesting the boundaries of the political*. Princeton University Press.
Habermas, J. (1999). The inclusion of the other: Studies in political theory. In C. Cronin & P. De Greiff (Eds.). Polity Press.
Habermas, J. (2006). *Time of transitions*. Edited and Translated by C. Cronin and Max Pensky. Polity Press.
Haidt, J. (2001). The emotional dog and the rational tail. *Psychological Review, 108*(4), 814–834.
Haldar, A. (2018a, November 2). Intrinsic goodness. *Times Literary Supplement.*
Haldar, A. (2018b, December 14). Economics: The discipline that refuses to change. *The Atlantic.*
Haldar, A. (2022a, February 15). Boris Johnson's last affair. *Project Syndicate*, https://www.project-syndicate.org/commentary/boris-johnson-partygate-scandal-by-antara-haldar-2022-02
Haldar, A. (2022b, April 18). Why do we obey the law? *Project Syndicate*, https://www.project-syndicate.org/commentary/ketanji-brown-jackson-bolsters-supreme-court-sagging-legitimacy-by-antara-haldar-2022-04
Haldar, A. (2022c, May 18). Losing rights in America. *Project Syndicate*, https://www.project-syndicate.org/commentary/supreme-court-reversing-womens-rights-by-antara-haldar-2022-05
Haldar, A. (2022d, June 29). Trump's lessons for defending the rule of law. *Project Syndicate*, https://www.project-syndicate.org/commentary/us-january-6-committee-trump-facts-emotion-by-antara-haldar-2022-06
Haldar, A. (2022e, October 26). Libertarian Lettuce. *Project Syndicate*, https://www.project-syndicate.org/commentary/truss-uk-libertarian-experiment-failed-by-antara-haldar-2022-10
Haldar, A., & Stiglitz, J. (2013). Analyzing legal formality and informality. In D. Kennedy & J. Stiglitz (Eds.), *Law and economics with Chinese characteristics*. Oxford University Press.
Hare, R. M. (1962). *Freedom and reason*. Oxford University Press.
Harman, G. (1988). Ethics and observation. In G. Sayre-McCord (Ed.), *Essays on moral realism*. Cornell University Press.
Hart, H. L. A. (1961). *The concept of law*. Clarendon Press.
Hauser, M. (2007). *Moral minds*. Little Brown.
Hegel, G. W. F. (1975). *Hegel's Aesthetics, lectures on fine art* (T. M. Knox, Trans.). Clarendon Press.
Heidegger M, 2009, *The Heidegger reader*, eds. Gunter Figal & Jerome Veith. Indiana University Press.
Hersch, S. M. (1972). *Cover-up: The Army's secret investigation of the massacre at My Lai 4*. Random House.
Holmes, A. F. (Ed.). (2005). *War and Christian ethics* (2nd ed.). Grand Rapids, Mich.
Holton, R. (1998). Positivism and the internal point of view. *Law and Philosophy, 17*(5/6), 597–625.
Homer (1900). *The Odyssey* (Butler, Trans.). AC Fifield.

Hookom, A. (2011). *But what kind of badness? An inquiry into the ethical significance of pain* (thesis). Georgia State University.
Huemer, M. (2005). *Ethical intuitionism*. Palgrave Macmillan.
Hume, D. (1888). In L. A. Selby-Bigge (Ed.), *Treatise of human nature*. Clarendon Press.
Hurka, T. (2001). *Virtue, vice and value*. Oxford University Press.
Jackson, F. (1998). *From metaphysics to ethics*. Oxford University Press.
Jackson, F., & Mitchell, D. B. (2007). *Philosophy of mind and cognition*. Blackwell Publishing.
James, W. (1884). What is emotion? *Mind, 9*(34), 188–205.
Jost, J., Banaji, M., & Nosek, B. (2004). A decade of system justification theory: Accumulated evidence of conscious and unconscious bolstering of the status quo. *Political Psychology, 25*(5), 881–919.
Joyce, R. (2001). *The myth of morality*. Cambridge University Press.
Kagan, S. (1992). The limits of well-being. In F. D. Miller Jr. & E. F. Paul (Eds.), *The good life and the human good*. Cambridge University Press.
Kagan, S. (1998). *Normative ethics*. Westview Press.
Kahane, G. (2011). Evolutionary debunking arguments. *Nous, 45*(1), 103–125.
Kahneman, D. (2000). Experienced utility and objective happiness: A moment-based approach. In D. Kahneman & A. Tversky (Eds.), *Choices, values and frames*. Cambridge University Press.
Kahneman, D., & Tversky, A. (1984). Choices, values, and frames. *American Psychologist, 39*(4), 341–350.
Kahnemann, D. (2011). *Thinking fast and slow*. Farrar, Straus and Giroux.
Kammerer, F. (2019). The normative challenge for illusionist views of consciousness. *Ergo, an Open Access Journal of Philosophy, 6*.
Kant, I. (1964). In H. J. Paton (Ed.), *Groundwork of the metaphysics of morals*. Harper & Row.
Katz, J. J., & Fodor, J. A. (1963). The structure of a semantic theory. *Language, 39*(2), 170–210.
Kauppinen, A. (2013). A Humean theory of moral intuition. *Canadian Journal of Philosophy, 43*(3), 360–381.
Kelly, T. (2005). The epistemic significance of disagreement. In J. Hawthorne & T. Szabo (Eds.), *Oxford studies in epistemology* (Vol. 1). Oxford University Press.
Kim, J. (2008). What is 'naturalized epistemology'. In E. Sosa & J. Kim (Eds.), *Epistemology: An anthology*. Blackwell Publishing.
Kinnier, R., Kernes, J., & Dautheribes, T. (2000). A short list of universal moral values. *Counseling and Values, 45*, 4–16.
Knobe, J., & Shapiro, S. (2021). Proximate cause explained: An essay in experimental jurisprudence. *The University of Chicago Law Review, 88*(1), 165.
Kohlberg, L. (1972). A cognitive-developmental approach to moral education. *The Humanist, 32*, 13–16.
Kohlberg L, ([1981, 1984]). *Essays on moral development*, vols. 1 and 2. Harper & Row.
Koonz, C. (2003). *The Nazi conscience*. Cambridge, Mass.
Lacan, J. (1973). *Les quatre concepts fondamentaux de la psychanalyse—livre XI*. Éditions du Seuil.
Lazarus, R. (1984). On the primacy of cognition. *American Psychologist, 39*, 124–129.
Legrand, P. (1997). On the impossibility of "legal transplants". *Maastricht Journal of European and Comparative Law, 4*(2), 111–124.
Lerner, A. (unpublished). Fine-tuning evolutionary debunking arguments.
Lieberman, D. E. (2013). *The story of the human body: Evolution, health, and disease*. Vintage.
Locke, J. (1690). *Essay concerning human understanding*. Bassett.
Lorenz, K. (1966). *On aggression*. Methuen.
Ludwig, K. A. (1994). First-person knowledge and authority. In G. Prayer, F. Siebelt, & A. Ulfig (Eds.), *Language, mind, and epistemology: On Donald Davidson's philosophy*. Springer Science Business Media: Kluwer Academic Publishers.
Ludwig, K. (2015). The sources of relativism. *Ethics, 126*, 175–195.
Lyotard, J.-F. (1971). *Discours, figure*. Klincksieck.

Lyotard, J. -F. (2011). *Discourse, figure* (A. Hudek & M. Lydon, Trans.). University of Minnesota Press.
Mackie, J. (1977). *Ethics: Inventing right and wrong*. Penguin.
MacMillan, M. (2014). *The war that ended peace: The road to 1914*. Random House.
Maldiney, H. (1973). *Regard, parole, espace*. Éditions L'Âge d'Homme.
Maldiney, H. (1985). *Art et existence*. Klincksieck.
Maldiney, H. (1993). *L'Art, l'éclair de l'être*. Les éditions du cerf.
Malinowski, B. (1931). Culture. In *Encyclopaedia of the social sciences* (Vol. 4). MacMillian.
Malraux, A. (1948). *La Création artistique*. Skira.
Markus, H. R., & Kitayama, S. (1991). Culture and the self: Implications for cognition, emotion and motivation. *Psychological Review, 98*(2), 224–253.
Markus, H. R., & Kitayama, S. (1994). The cultural shaping of emotion: A conceptual framework. In S. Kitayama & H. R. Markus (Eds.), *Emotion and culture: Empirical studies of mutual influence* (pp. 339–351). American Psychological Association.
Maroney, T. (2016). A field evolves: Introduction to the special section on law and emotion. *Emotion Review, 8*(1), 3–7.
Marrin, A. (1974). *The last crusade: The Church of England in the first world war*. Duke University Press.
McCloskey, D. N. (2006). *The bourgeois virtues: Ethics for an age of commerce*. University of Chicago Press.
McCloskey, D. N. (2019). *Why liberalism works: How true Liberal values produce a freer, more equal, prosperous world for all*. Yale University Press.
McGinn, C. (1996). *The character of mind*. Oxford University Press.
McGinn, C. (2017). *The philosophical provocations*. MIT Press.
Mead, M. (Ed.). (1959). *An anthropologist at work: Writings of Ruth Benedict*. Houghton Mifflin.
Mendola, J. (1990). Objective value and subjective states. *Philosophy and Phenomenological Research, 50*(4), 695–713.
Mendola, J. (2006). Intuitive hedonism. *Philosophical Studies, 128*, 441–477.
Merleau-Ponty, M. (1949). Le doute de Cézanne. In *Sens et non-sens* (pp. 15–49). Les Éditions Nagel.
Merleau-Ponty, M. (1964). *L'Œil et l'esprit*. Gallimard.
Merli, D. (2008). Expressivism and the limits of moral disagreement. *The Journal of Ethics, 12*, 25–55.
Moore, G. E. (1903). *Principia Ethica*. Cambridge University Press.
Nagel, T. (1995). Wittgenstein: The egocentric predicament. In *Other minds: Critical essays 1969–1994*. Oxford University Press.
Narveson, J. (1994). The agreement to keep our agreements: Hume, Prichard, and Searle. *Philosophical Papers, 23*(2), 75–87.
Narveson, J. (2010). The relevance of decision and game theory to ethical theory. *Ethical Theory and Moral Practice, 13*(5), 497–520.
Narveson, J. (2016). Social contract: The only game in town. *Dialogue, 55*(4), 695–711.
Nietzsche, F. (1998). *On the genealogy of morality* (M. Clark & A. J. Swensen, Trans.). Hackett.
Nietzsche, F. (2002). *Beyond good and evil* (J. Norman, Trans.). Cambridge University Press.
Nozick, R. (1974). *Anarchy state and utopia*. Basic Books.
Nussbaum, M. (2015). *Political emotions: Why love is necessary for justice*. Harvard University Press.
Perry, S. R. (1995). Interpretation and methodology in legal theory. *Law and Interpretation, 97*(99), 188–189.
Pfaff, D. (2007). *The neuroscience of fair play*. Dana.
Pinker, S. (2011). *The better angels of our nature: Why violence has declined*. Viking.
Pistor, K., Haldar, A., & Amirapu, A. (2009). Social norms, rule of law and gender reality. In J. J. Heckman, R. Nelson, & L. Cabbatingan (Eds.), *Global perspectives on the rule of law*. Routledge-Cavendish.

Popper, K. (1994). *Knowledge and the body-mind problem.* Routledge.
Postema, G. J. (1998). Jurisprudence as practical philosophy. *Legal Theory, 329*, 332.
Putnam, H. (1986). Other minds. In *Mind, language and reality* (Vol. 2). Cambridge University Press.
Quine, W. V. (1969). Epistemology naturalized. In *Ontological relativity and other essays.* Columbia University Press.
Quine, W. V. (1977). Mind and verbal dispositions. In S. Guttenplan (Ed.), *Mind and language.* Clarendon Press.
Railton, P. (1986). Moral realism. *Philosophical Review, 95*(2), 163–207.
Railton, P. (1989). Naturalism and prescriptivity. *Social Philosophy and Policy, 7*(1), 151–174.
Raz, J. (1979). *The authority of law: Essays on law and morality.* Clarendon Press.
Reich, D. (2018). *Who we are and how we got here: Ancient DNA and the new science of the human race.* Pantheon.
Rhees, R. (1971). Could language be invented by a Robinson Crusoe. In O. R. Jones (Ed.), *The private language argument.* Macmillan St Martin's Press.
Ricœur, P. (1965). *De l'interprétation. Essai sur Freud.* Éditions du Seuil.
Ricœur, P. (1969). *Le conflit des interprétations. Essais d'herméneutique.* Le Seuil.
Ricoeur, P. (1974). *The conflict of interpretations: Essays in hermeneutics* (W. Domingo et al., Trans., D. Ihde, Ed.). Northwestern University Press.
Rimland, B. (1982). The altruism paradox. *Psychological Reports, 51*, 221–222.
Robbins, J. (2018, April 26). *Native knowledge: What ecologists are learning from indigenous people.* Yale Environment 360. Yale University: online journal published by the Yale School of Forestry and Environmental Studies.
Rovane, C. (2013). *The metaphysics and ethics of relativism.* Harvard University Press.
Rovane, C. (2015, November 23). Moral dispute or cultural difference. *The New York Times.*
Sayre-McCord, G. (1988). Moral theory and explanatory impotence. In G. Sayre-McCord (Ed.), *Essays on moral realism.* Cornell University Press.
Sazaklis, J., & Luciano, V. (2015). *Wonder woman: An origin story.* Camstone.
Schauer, F. (2015). *The force of law.* Harvard University Press.
Scheuerman, W. (2017, April 12). *Habermas and the fate of democracy.* Boston Review.
Schnall, S. (2008). Disgust as embodied moral judgment. *Personality and Social Psychology Bulletin, 34*(8), 1096–1109.
Schroeder, M. (2008). *Being for.* Oxford University Press.
Schroeder, M. (2021). A common subject for ethics. *Mind, 130*(517), 85–110.
Schwitzgebel, E. (2008). The unreliability of naive introspection. *Philosophical Review, 117*(2), 245–274.
Shapiro, S. (2006). What is the internal point of view? *Fordham Law Review, 75*(3), 1157.
Shaver, R. (2004). The appeal of utilitarianism. *Utilitas, 16*(3), 235–250.
Sheiman, B. (2009). *An atheist defends religion.* Penguin.
Sinhababu, N. (2017). *Humean nature.* Oxford University Press.
Sinhababu, N. (forthcoming-a). Naturalistic arguments for ethical hedonism. In R. Y. Chappell, D. Meissner, & W. MacAskill (eds.), *An introduction to Utilitarianism.*
Sinhababu, N. (forthcoming-b). The reliable route from nonmoral evidence to moral conclusions. *Erkenntnis* 1–21.
Smith, M. (1994). *The moral problem.* Wiley-Blackwell.
Smith, M. (2021). The state of nature. In M. Timmons (Ed.), *Oxford studies in normative ethics* (pp. 270–289). Oxford University Press.
Smuts, A. (2011). The feels good theory of pleasure. *Philosophical Studies, 155*, 2.
Sobel, D. (2019). The case for stance-dependent reasons. *Journal of Ethics and Social Philosophy, 15*, 146.
Sommers, Roseanna, (2021): "Experimental jurisprudence", Science, Vol. 373, Issue 6553, pg. 394–395.
Stewart, T. (2005). The competing practices argument and self-defeat. *Episteme*, 13–24.
Strawson, P. (1977). *Individuals.* Methuen.

Street, S. (2006). A Darwinian dilemma for realist theories of value. *Philosophical Studies, 127*(1), 109–166.
Susen, S. (2017). *Jurgen Habermas from the Routledge handbook of language and politics*. Routledge. https://doi.org/10.4324/9781315183718.ch3
Tannsjo, T. (2007). Narrow hedonism. *Journal of Happiness Studies, 8*, 79–98.
Tappolet, C. (2016). *Emotions, values, and agency*. Oxford University Press.
Taylor, C. (1975). *Hegel*. Cambridge University Press.
Taylor, C. (1994). The politics of recognition. In A. Gutmann (Ed.), *Multiculturism: Charles Taylor* (pp. 25–75). Princeton University Press.
Thaler, R., & Sunstein, C. (2008). *Nudge: Improving decisions about health, wealth and happiness*. Yale University Press.
Thich, N. H. (1999). *The heart of the Buddha's teaching*. Three River Press.
Thomson, J. (2008). *Normativity*. Open Court.
Thornton, T. (1998). *Wittgenstein on language and thought: The philosophy of content*. Edinburgh University Press.
Tobia, K. (2022). Experimental jurisprudence. *The University of Chicago Law Review, 89*(3), 735–802.
Tsai, J., & Clobert, M. (2019). Cultural influences on emotion: Empirical patterns and emerging trends. In S. Kitayama & D. Cohen (Eds.), *Handbook of cultural psychology* (pp. 292–318). Guilford Press.
Tuttle, R. H. (2014). *Apes and human evolution*. Harvard University Press.
Vavova, K. (2014). Moral disagreement and moral skepticism. *Philosophical Perspectives, 28*, 302–333.
Warnasuriya, K. S. (2007). Dependent origination and the Buddhist theory of relativity. *Hsi Lai Journal of Humanistic Buddhism (西來人間佛教學報), 8*, 153–163.
Watson, A. (1974). *Legal transplants: An approach to comparative law*. Scottish Academic Press.
Werner, P. (2014). Moral perception and the contents of experience. *Journal of Moral Philosophy*, 1–24.
White, F. J., Waller, M. T., & Boose, K. J. (2013). Evolution of primate peace. In D. P. Fry (Ed.), *War, peace, and human nature: The convergence of evolutionary and cultural views* (pp. 389–405). Oxford University Press.
Williams, B. (1972). *Morality: An introduction to ethics*. Cambridge University Press.
Wilson, M. L. (2013). Chimpanzees, warfare, and the invention of peace. In D. P. Fry (Ed.), *War, peace, and human nature: The convergence of evolutionary and cultural views* (pp. 361–388). Oxford University Press.
Wittgenstein, L. (1953). *Philosophical investigations*. Blackwell.
Wittgenstein, L. (1994). *The Wittgenstein reader*, ed. Anthony Kenny. Blackwell Publishing.
Wren-Lewis, S. (2015). Towards a widely acceptable framework for the study of personal well-being. In *Well-being in contemporary society* (pp. 17–38).
Zajonc, R. (1984). On the primacy of affect. *American Psychologist, 39*, 117–123.
Zhuangzi, Z. (2009). *The essential writings* (B. Ziporyn, Trans.). Hackett Publishing Company.
Zweigert, K., Kötz, H., & Weir, J. A. (1987). *An introduction to comparative law* (2nd edn).

MIX
Papier aus verantwortungsvollen Quellen
Paper from responsible sources
FSC® C105338

If you have any concerns about our products,
you can contact us on
ProductSafety@springernature.com

In case Publisher is established outside the EU,
the EU authorized representative is:
Springer Nature Customer Service Center GmbH
Europaplatz 3, 69115 Heidelberg, Germany

Printed by Libri Plureos GmbH
in Hamburg, Germany